THE EVOLVING BRAIN
The Mind and the Neural Control of Behavior

THE EVOLVING BRAIN
The Mind and the Neural Control of Behavior

by

C. H. Vanderwolf
University of Western Ontario
London, Ontario, Canada

 Springer

ISBN-13: 978-1-4419-4170-1
e-ISBN-13: 978-0-387-34230-6

Printed on acid-free paper.

9 8 7 6 5 4 3 2 1

springer.com

Contents

Preface

The study of the higher level neural control of behavior has been dominated by the theory that many aspects of cerebral activity are functionally organized in accordance with psychological concepts such as perception, attention, motivation, memory, emotion or cognition. I believe that this entire approach is misguided because it is based on false assumptions derived from the speculations of the ancient Greek philosophers. The series of essays in this book discusses the implications of a mentalistic approach to the study of brain function and points out the absence of significant progress associated with it. The alternative that is proposed is that we abandon attempts to discover the neural basis of mind as classically conceived and turn instead to an analysis of the neural mechanisms that control behavior. This broad topic touches on a variety of traditional fields. Therefore, the material discussed in this book may be of interest, not only to neuroscientists and psychologists, but also to animal behaviorists, anthropologists, evolutionary biologists, neurologists, philosophers, psychiatrists, and others interested in the general field of the brain, behavior and the mind.

Acknowledgements

I am indebted to the University of Western Ontario which provided financial support for the preparation and publication of this book; to Daniella Chirila for her patience in typing the manuscript; and to Francis Boon who prepared the figures. I am also indebted to: Dr. Lee Foote (University of Alberta, Edmonton, Alberta) for helpful comments on Chapter IV; and to Dr. Martin Kavaliers (University of Western Ontario, London, Ontario) and Dr. T.E. Robinson (University of Michigan, Ann Arbor, Michigan), who pointed out some useful references to me.

I. The mind and the explanation of behavior

It is conventional to explain human behavior in terms of mental activity. We are said to act as we do because of desires, wishes, opinions, beliefs, motives, etc. This common sense approach to the mind and behavior has been very influential in the broad field of brain research and neuroscience. In the past half century an enormous research effort has been devoted to the study of the neural basis of cognition (cognitive science, cognitive neuroscience), of memory, and also of attention, motivation and emotion. It appears to be widely assumed that we are in possession of a valid taxonomy of mental processes, a fund of well-established knowledge about the organization of high level neural activity that is obvious to everyone. What is the nature of this taxonomy, how was it established and agreed on, and lastly, can we be certain of its validity?

Present day ideas about the mind do not appear to have departed very far from the classic summary of psychological knowledge provided by William James in 1890.[1] Chapter headings listed by James include: "The stream of thought, The consciousness of self, Attention, Conception, Discrimination and comparison, Association, The perception of time, Memory, Sensation, Imagination, The perception of things, The perception of space, The perception of reality, Reasoning, Instinct, The emotions, and Will".

David Hume, writing in the 18th century[2] provided a similar though more extensive list of mental faculties, processes, or states including the following: "impressions, ideas, pride, humility, pleasure, pain, vice, virtue, vanity, wit, humour, love of fame, sentiments, passions, love, hatred, esteem for the rich and powerful, sympathy, benevolence, anger, compassion, pity, malice, envy, respect, contempt, amorous passion, desire, aversion, grief, joy, hope, fear, will, imagination, curiosity, reason, understanding, moral sense, feelings, selfishness, generosity, a sense of justice, beliefs, respect, vanity, prejudice, gratitude, zeal, disinterestedness, fidelity, esteem, industry, perseverance, patience, vigilance, application, constancy, temperance, frugality, irresolution, uncertainty, reveries, thoughts."

In addition to all the foregoing, one cannot ignore such concepts as the conscious mind, the preconscious, the unconscious, the ego, the id, the superego, repression, and sublimation. All these concepts, and more, were introduced by Sigmund Freud in the 20th century.[3]

If one seeks the source of this long mentalistic tradition in the history of Western thought, one comes, at last, to Aristotle, a Greek philosopher living from 384–322 BC,[4] and his teacher Plato (428–348 BC). Aristotle proposed that living things differ from non-living things because they possess a non-corporeal psyche. The presence of the psyche, he thought, keeps the body together throughout a long life but at death, when the psyche has departed, the body speedily rots and disintegrates (especially in a hot Greek summer!). All living things, said Aristotle, possess a vegetative psyche responsible for nutrition, growth and reproduction. Plants, he said, have no further psychic powers but animals have both a vegetative psyche and a sensitive psyche, permitting reactivity to touch and other sensory stimuli. Only humans possess the highest type of psyche which confers a capacity for rational thought. In addition to these major subdivisions, the Aristotelian psyche also possessed numerous faculties such as desire, opinion, memory, imagination, belief, judgment, conviction, thinking, etc. Aristotle's theories of the psyche and of many other topics in what we now regard as physics, chemistry and biology were adopted by the Christian Church and disseminated throughout the Western world over a period of many centuries.[5] As a result his ideas were widely accepted. However, the discovery by William Harvey (1578–1657) that the circulation of the blood is a mechanical process and later work such as the discovery by Antoine Lavoisier (1743–1794) that animal heat and life depend on chemical processes gradually led to a general acceptance of the idea that life processes are dependent on physical and chemical processes. The Aristotelian theory of a psyche that was responsible for the phenomena of life became unnecessary.

It appears that the French philosopher Rene Descartes (1596–1650) played a major role in establishing the mechanistic point of view in biology.[6,7] Descartes assumed that the bodies of humans and all aspects of the functioning of non-human animals depended on mechanical principles. Animal behavior was attributed to reflexes, simple sensori-motor reactions involving the nervous system, but human behavior, although partly reflex, was held to be mainly dependent on the activity of a rational soul. These ideas had two important effects: (a) the study of the function of the body, up to and including the level of reflexes, could be studied freely by physical and chemical methods, giving rise to modern physiological science; and (b) human behavior was placed outside the field of materialistic science, effectively separating psychology from the rest of biological science and permitting Aristotelian ideas about the higher levels of the psyche to persist into modern times.

To a modern scientifically literate reader, most of Aristotle's ideas seem bizarre and primitive. He tells us that circular motion is the fundamental type but Galileo and Newton taught us to think that linear motion is fundamental. Aristotle thought that falling objects move at a constant velocity; having no understanding whatever of gravity, he did not realize that falling bodies accelerate. Knowing nothing about chemistry, Aristotle accepted the theory that all material objects are made up of four elements: fire, water, earth and air. We recognise a periodic table listing up to 107 elements that have no resemblance to Aristotle's elements.

In contrast, Aristotle's discussion of psychological topics sounds rather modern. Reason is said to be distinct from emotion, and is often opposed to it. Thought always involves mental images and thought proceeds by a process of association of ideas. Memory is compared to a physical information storage device (a signet ring pressed into wax) in a manner that has many parallels with modern comparisons of human memory to computer memory. There can be little doubt that although Aristotelian ideas have been supplanted in physics, chemistry and biology they have persisted to the present in philosophy, psychology, psychiatry and common popular opinion.

As an example of the process by which mentalistic concepts were developed, let us consider the origin of the concept of cognition which forms the intellectual basis of present-day cognitive science and cognitive neuroscience. In the *Republic*, Plato[8] concludes that the ideal state should consist of three social classes: a) rulers; b) soldiers; and c) farmers and workers of all kinds. Further, Plato thought, what is true of the state must also be true of individuals. Therefore, the psyche will also consist of three parts: a) reason, intellect or cognition (corresponding to the rulers); b) feelings, spirit, will or conation (corresponding to the soldiers); and c) desires, emotions or appetites (corresponding to the farmers and workers). As evidence favouring this tripartite division of the psyche, Plato pointed to the common observation that people often seem to experience internal conflicts. For example, a man might be thirsty yet unwilling to drink.

Although conation is rather neglected nowadays, cognition and emotion figure prominently in cognitive neuroscience and the philosophy of mind. It is, for example, widely believed that there is a separate entity, the limbic system of the brain, which is the basis for emotion while the neocortex and its connections provide the basis for intellect or cognition. However, one may legitimately ask whether Plato and his followers really got it right. Are reason, cognition, etc., really different in principle from desires, emotions, appetites, etc.? When making decisions in everyday life, people often seem to have difficulty distinguishing among self-interest, prejudice, and a logical consideration of the available evidence. If such things were truly different there should be no such difficulty. Self-deception would be less common than it is now. If a thirsty man does not drink, perhaps because he thinks the available water may be contaminated, one need not assume a conflict between desire and intellect, as Plato thought. Perhaps there is a conflict between two desires (thirst versus a desire to avoid illness). Perhaps there is a conflict between two equally rational ideas: a) this water will do me good; and b) this water will do me harm. Are arguments and evidence of the type presented by Plato really sufficient to decide the question of the overall organization of the mind or the brain? Is it reasonable to lump together such diverse things as hunger, thirst, fear, rage, hatred and sexual lust into a single category? Why should Plato's idea of a tripartite psyche be taken seriously?

It is widely believed that the conventional theory of the mind or psyche can be verified by simple introspective examination of one's own thoughts, feelings, motives, etc. Rene Descartes wrote: "I see clearly that there is nothing which is easier for me to know than my mind."[7] However, a systematic attempt to analyze the mind in detail by introspection in the period between approximately 1880–1910 lead to failure and the conclusion that introspection is not a valid method of study.[9] What one might call "mentation" or "cerebration" is generally not available to introspection. There is a good deal of evidence that what people are really aware of when they "introspect" is sensory input from muscles, joints, viscera, etc.[10] There appears to be no capacity for the mind to examine itself directly. The conventional sensory channels (visual, auditory, gustatory, olfactory, tactile, thermoceptive, proprioceptive, nociceptive, and interoceptive inputs) provide information about the state of the body and the outside world, not the mind or the brain. Therefore, the conventional taxonomy of mental processes cannot be verified by "introspection".

The conclusion that introspection is impossible, that one cannot directly observe one's own mental activity, is intuitively implausible. As William James put it (1, p. 185) "The word introspection need hardly be defined – it means, of course, the looking into our own minds and reporting what we there discover". If we live a life of comfortable routine, we know very well our own likes and dislikes and we feel confident that we know what we will do in the future. Surely, a critical reader may suggest, this is due to introspection. Doubts about this may appear if the settled routine of everyday life is suddenly overthrown and one finds one's self unexpectedly in great physical danger or in any situation that elicits a strong reaction, violent sexual jealousy, for example. One reacts to such situations in ways that may, on later sober reflection, appear admirable or shameful, but in all such cases it seems to be common to be rather startled by one's own behavior. One asks: "How could I have done that?"

It may be that we are familiar with our own behavior, not through any direct insight into the mechanisms that cause that behavior, but merely because we have, many times over, experienced the sensory consequences of that behavior in the past. Formal evidence that this is indeed the case is provided by a famous series of experiments on obedience to authority by Stanley Milgram of Yale University.[11] Under the guise of an experiment on the effect of punishment on human learning, naïve subjects were instructed to deliver electric shocks to a man strapped in a chair (the victim) whenever the victim made an error in a learning task. Although severe shocks were never, in fact, applied, the naïve subject was lead to believe that he was administering shocks of increasing intensity up to a level that might be dangerous (450 volts). Under the various conditions of the experiment, 30–65% of the naïve subjects were willing to administer shocks at the maximum voltage even though the victim, apparently a talented actor, was struggling and screaming, and even though, under one condition, the naïve subjects had to hold the victim's hand forcibly on the shock plate. Thus, a high proportion of normal adult men will obey an authority (the experimenter) who orders them to do cruel and dangerous things to other people.

These results, in addition to their relevance to the question of how despotic regimes can induce ordinary people to perform acts of torture and murder, have relevance to the question of how well people know their own mind. Milgram asked groups of people (college students, middle-class adults) who had not actually taken part in these experiments but had the methods used described to them, how they themselves would have reacted if they had played the role of naïve subjects. Not one of a group of 110 people believed themselves willing to deliver high intensity shocks to the victim. A group of 39 psychiatrists thought that perhaps one person in a thousand (0.1%) would be willing to do it, not the 30–65% that actually will do it. We can conclude that people have no introspective access to the behavioral control mechanisms that are activated by the commands of someone in authority.

There is also reason to doubt that humans have conscious access to the mechanisms that control purposive behavior in a general sense. It is conventional to believe that people do things that result in a feeling of pleasure and avoid doing things that result in pain. A clear demonstration that this may not be entirely correct is provided by an experiment on the reinforcing and subjective effects of morphine administration in men with a past history of intravenous morphine use (post-addicts).[12] The term "reinforcing effect" refers here to the ability of morphine injections to increase the rate of pressing a lever above the rate obtainable with control (placebo) injections if, and only if, the morphine injections are dependent on pressing the lever. The term "subjective effects" refers here to the ability of the post-addicts to demonstrate that they could detect the morphine injection by correctly stating, on a questionnaire, that they had received the morphine and not the placebo. A dose of morphine of 3.75 mg maintained lever pressing above control levels in four of the five post-addicts, and doses of 7.5, 15 and 30 mg maintained lever pressing in all five cases. However, according to the questionnaire results, the post-addicts were aware only of the 30 mg dose. These results show that the reinforcing effect of morphine is not dependent on a pleasurable effect that can be reported verbally (on a questionnaire). This is consistent with the general conclusion that behavior control mechanisms are not open to introspective examination. We behave as we do as a result of the properties of the neural circuitry controlling behavior and not as a result of any subjective feelings we may have.

In addition to the apparent non-existence of genuine introspection, there is another reason for doubting the validity of the mentalistic concepts bequeathed to us by the philosophers of ancient Greece. Western and non-Western civilizations have devised different psychological systems. This point was demonstrated quite clearly in a book by K. Danziger, a Canadian psychologist who spent two years teaching at a university in Indonesia.[13] When Danziger discovered that the host university already had a type of psychologist whose teachings were based on Hindu philosophy, he suggested that the two of them organize a joint seminar in which Eastern and Western approaches to psychological problems could be compared. However, when he suggested potential seminar topics such as learning, motivation or intelligence, the Indonesian objected that the findings Danziger wished to include under each of these headings were heterogenous collections of phenomena that had nothing interesting in common. Conversely, the topics suggested by the Indonesian appeared incomprehensible to Danziger. Since it proved to be impossible to agree on suitable topics for discussion, the proposed joint seminar never took place.

The difficulties experienced by K. Danziger and his Indonesian colleague suggest that the familiar concepts of conventional psychology are purely verbal constructs, useful in human discourse but having no real biological validity. In much the same way we can speak of "learning by heart," "affairs of the heart", having a "hard heart", a "soft heart," or a "broken heart" without implying any relation to the hollow muscular organ that contracts rhythmically in every human thorax. Expressions of this type, persisting in everyday speech, are another indication of the persisting influence of Aristotelian ideas: Aristotle thought that the various components of the psyche were associated particularly with the heart.

A final reason for doubting the validity of the conventional theory of the mind is that it has not been very successful in stimulating new discoveries. Numerous authors have pointed out that no major advances have been made in psychology in a long period despite a prodigious amount of research activity.[14] A similar situation prevails in much of behavioral or cognitive neuroscience. For example, the theory that there is a localized region of the brain, the hippocampus, which is responsible for the conventional faculty of memory enjoyed almost universal support for over 40 years but is now no longer regarded as valid by a growing number of investigators.[15] There is serious doubt that "memory" is a meaningful functional category of brain activity in the same sense that "visual activity" or "auditory activity" are meaningful categories. The functions of the different sensory systems are anatomically localized by the existence of specialized receptors and their connections to the nervous system but there is no good reason to think that conventional psychological functions are localized in the same way.

As a further example of the failure of mentalistic approaches to neuroscience, studies of the neural basis of attention have led to no definite advance but only to a jumble of proposals relating this presumed mental entity or process to: (a) the thalamic intralaminar nuclei and the brain stem reticular formation; (b) the hippocampus; (c) the cingulate cortex; (d) the frontal cortex; (e) the parietal cortex; (f) the cholinergic projections from the basal forebrain to the neocortex; (g) noradrenergic projections from the locus coeruleus to the cerebral cortex; (h) long-term potentiation in the entorhinal projections to the dentate gyrus and Ammon's horn; (i) the pyriform cortex; (j) peripheral filtering of non-attended inputs; and (k) a miscellaneous group of structures including the amygdala, globus pallidus, and superior colliculus.[16] There is no scientific advance in any of this: it remains merely a mass of conflicting speculative proposals which have been neither refuted nor strongly supported.

Cognitive neuroscience is currently in the midst of a grand program of applying the new brain imaging technologies to the study of mental processes as classically conceived. If the arguments advanced here are valid, we can expect that this program will result in: (a) a modest amount of new knowledge about the location of various sensori-motor processes in the human brain; and (b) a mass of contradictory and inconclusive data, leading to disillusionment and abandonment of the original program.

A major problem in attempts to study the conscious mind is that no one has been able to devise a certain method for determining the presence or absence of subjective experience in other living (or non-living) things. If we cannot decide when subjective experience is present and when it is not it is impossible to determine what its physical basis might be. Descartes proposed that subjective experience is present only in living things that possess: (a) intelligent speech; and (b) reason, i.e. genuine understanding. Therefore, according to him, humans have subjective experiences but animals do not. Some of Descartes' followers put this doctrine into practice, as shown in the following quotation. "They administered beatings to dogs with perfect indifference, and made fun of those who pitied the creatures as if they had felt pain. They said that the animals were clocks; that the cries they emitted when struck were only the noise of a little spring which had been touched, but that the whole body was without feeling. They nailed poor animals up on boards by their four paws to vivisect them and see the circulation of the blood which was a great subject of conversation."[17]

Subsequent opinion has rejected Descartes' proposal that animals, especially non-human mammals, can be regarded as automata completely devoid of subjective experience. There are probably very few people alive today who believe that dogs cannot feel pain even though they cannot speak as humans do and may be rather deficient in reasoning abilities. However, an essentially Cartesian distinction between sentient and non-sentient neural structures is very widely accepted. Thus, modern discussions of reflex responses in the isolated mammalian spinal cord are restricted to the physico-chemical processes occurring in neurons; no one ever discusses the possible presence of subjective experiences in the spinal cord. Yet this was not always the case. Edward Pflüger, an eminent 19th century physiologist, proposed the existence of a spinal consciousness.[18] Various other authors have proposed the existence of subjective experience in insects and in micro-organisms (protozoa, bacteria).[19] A recent scholarly paper, reviving the ancient hypothesis of pan-psychism, has proposed that subjectivity is a property of virtual photons, leading to the conclusion "that the whole universe must be imbued with subjectivity".[20]

Such questions are not merely arcane academic matters. A knowledge of the extent to which various species can experience pain and suffering would contribute greatly to the humane treatment of animals. The problem of determining the presence of subjective experience assumes great practical importance in medicine in the condition known as the "locked-in syndrome". After recovery from anesthesia, surgical patients sometimes complain of having suffered intense pain during the procedure even though they could not speak or move at the time and appeared to the anesthetist to be fully anesthetized. Such reports are often accompanied by accurate descriptions of events occurring during the surgical procedure, such as a detailed account of conversations among the surgical team members. Therefore, claims of preserved consciousness during a state of what outwardly appears to be surgical anesthesia cannot be dismissed as due to false memories or hallucinations.[21]

In one case[22] a woman who had been judged to have totally lost the capacity for consciousness after a severe head injury eventually recovered and informed the world that she had been fully conscious even while plans were underfoot to remove her life support systems and allow her to die. Such cases demonstrate that conscientious trained professionals have sometimes failed to detect the presence of subjectivity when it seems to have been present.

Related problems can occur after localized brain damage. After section of the forebrain commissures (mainly the anterior commissure and the corpus callosum) a neurosurgical patient may be able to name common household objects concealed in a bag after feeling them with the right hand. This is possible because somatosensory information from the right hand can reach the left cerebral hemisphere in which the speech areas are usually located. However if the objects are felt with the left hand, the patient cannot name the objects because somatosensory information cannot reach the left hemisphere, but may be able to reveal a knowledge of their uses by demonstrations with the left hand which is controlled by the right hemisphere. According to R.W. Sperry[23] the main discoverer of these intriguing phenomena, both the right and left hemispheres in such patients are fully conscious even though only one hemisphere is capable of speech. However, according to J.C. Eccles,[24] evidently an implicit believer in the Cartesian criterion of speech as an infallible indicator of consciousness, subjective experience is entirely confined to the speaking hemisphere. Right hemisphere activity, Eccles tells us, can become conscious only after transmission to the left (speaking) hemisphere via intact forebrain commissures.

An implicit acceptance of speech as the highroad to conscious experience is also apparent in discussions of "blindsight", a condition associated with striate cortex lesions, in which patients may continue to demonstrate some visually guided behaviors (such as accurate reaching for objects) but verbally deny that they can see those same objects.[25] This phenomenon demonstrates that the striate cortex is the site of visual consciousness only if one assumes that an absence of relevant speech is a certain indicator of a lack of consciousness.

A conceptually related phenomenon occurred in the case of a young woman (D.F.) who suffered a localized bilateral occipital brain lesion as a result of carbon monoxide poisoning.[26] This injury had no effect on many visuomotor abilities such as the ability to step over sticks or rocks while walking or ability to orient the position of the hand and adjust the size of the grasp appropriately when picking up objects. Despite this, the patient could not verbally describe the orientation or size of objects and could not indicate this information by gestures.

Anatomical data provide a possible interpretation of these phenomena. It appears that there are two cortico-cortical output pathways from the visual cortex in the occipital lobe. A dorsal pathway to the parietal lobe, intact in D.F., appears to be responsible for a variety of visuomotor abilities. A ventral pathway to the temporal lobe, severely damaged in D.F., appears to be responsible for visual activation of human communication abilities. The speech areas of the dominant hemisphere in humans are involved in both vocal speaking and in gestures such as the manual sign language of the deaf.[27] In the patient D.F. the brain circuits involved in communication cannot be activated by visual stimuli, but brain circuits involved in controlling locomotion and manipulation can still be activated in this way.

The foregoing interpretation of the symptoms present in D.F., however, is not the one offered by Goodale and Milner who discussed the case in detail in a recent book.[26] Goodale and Milner instead propose that the dorsal occipito-parietal pathway activates unconscious actions while the ventral occipito-temporal pathway activates conscious perceptions. The grounds for denying consciousness to the dorsal pathway are similar to the grounds used by Eccles for denying consciousness to the minor hemisphere in patients with transection of the forebrain commissures; i.e., an absence of control of speech. This is the type of argument used by Descartes and his followers to deny consciousness to dogs and other non-human animals. Why should we accept a Cartesian argument in the one case and not in the other?

Conclusions. I suggest that the following conclusions can be drawn from these various facts and arguments. (1) Most of the conventional beliefs about the mind are based, not on factual evidence, but on ancient speculative philosophical theories. (2) There is, at present, no clear objective means of establishing the existence of subjective experience outside of one's self. This poses an enormous problem for any attempt to investigate the physical basis of such experience. (3) The mechanisms that control behavior are, in general, not open to introspective analysis. (4) Since there is very little reliable evidence concerning the nature of mind or the conditions necessary for its existence, conventional beliefs about the mind are not a valid basis for any program of investigation of the functional organization of the brain.

Notes

1. James, W. (1950). *The principles of psychology*, New York: Dover Publications (first published 1890).
2. Hume, D. (1978). *A treatise of human nature*, Oxford: Clarendon Press (first published 1739–40).
3. Freud, S. (1933). *New introductory lectures on psycho-analysis*. New York: W.W. Norton and Company.
4. Barnes, J. (1984). *The complete works of Aristotle, volumes 1 and 2.* Princeton, N.J., Princeton University Press.
5. A discussion of Aristotle's influence on Christian thought can be found in: Russell, B. (1961). *History of western philosophy*. London: Allen and Unwin. Also see: Magoun, H.W. (1958). Early development of ideas relating the mind with the brain. In: Wolstenholme, G.E.W., and O'Connor, C.M. (eds.) *Neurolgical basis of behavior*, Ciba Foundation Symposium, London: Churchill, pp. 4–27.
6. Huxley, T.H. (1970). *Collected essays* (1893–1894) *volume 1, Method and results.* Hildesheim: Georg Olms.
   Smith, H.W. (1959). The biology of consciousness. In: C.M. Brooks and P.F. Cranefield (eds). *The historical development of physiological thought.* New York: Hafner, pp. 110–136.
7. Haldane, E.S. and Ross, G.R.T. (1955). *The philosophical works of Descartes*: Volume 1. New York: Dover publications (First published by Cambridge University Press, 1911). Reprinted with corrections in 1931.
8. Demos, R. (1939). *The philosophy of Plato*, New York: Charles Scribner's Sons, p. 92.
   Grube, G.M.A. (1935). *Plato's thought*, London: Methuen and Co., pp. 120–149.
9. Boring, E.G. (1953). A history of introspection. *Psychological Bulletin, 50*: 169–189.
   Hebb, D.O. (1980). *Essay on mind.* Hillsdale, N.J. Lawrence Erlbaum.
   Hebb, D.O. (1977). To know your own mind. In: J.M. Nicholas (ed.) *Images perception and knowledge.* Dordrecht: Reidel; pp. 213–219.
   Humphrey, G. (1951). *Thinking: an introduction to its experimental psychology.* New York: Wiley.
   Lyons, W. (1986). *The disappearance of introspection.* Cambridge, Mass: MIT Press.

Nisbett, R.E., and Wilson, T.D. (1977). Telling more than we can know: verbal reports on mental processes. *Psychological Review, 84*: 231–259.
10. Vanderwolf, C.H. (1998). Brain, behavior, and mind: What do we know and what can we know? *Neuroscience and Biobehavioral Reviews, 22*: 125–142.
11. Milgram, S. (1974). *Obedience to authority: an experimental view*, New York: Harper and Row.
12. Lamb, R.J., Preston, K.L., Schindler, C.W., Meisch, R.A., Davis, F., Katz, J.L., Henningfield, J.E., and Goldberg, S.R. (1991). The reinforcing and subjective effects of morphine in post-addicts: a dose-response study. *Journal of Pharmacology and Experimental Therapeutics, 259*: 1165–1173.
In order to receive an injection, the post-addicts had been trained on a fixed ratio-100 response schedule. This means that 100 lever presses were followed by turning on a red light for 1.0 seconds. When 30 such fixed-ratio responses (a total of 3,000 lever presses) had been completed, the light came on for 15 minutes and morphine or placebo was administered intramuscularly. Each drug condition was in force for one week and neither the experimenters nor the post-addicts knew what was in the syringe (double blind design). The number of correct responses on the questionnaire were: 38% for the 3.75 mg dose; 59% for the 7.5 mg dose, 44% for the 15 mg dose; and 98% for the 30 mg dose (by chance alone one would expect correct responses about 50% of the time).
13. Danziger, K. (1997). *Naming the mind: how psychology found its language.* London: Sage Publications.
14. Lykken, D.T. (1991). What's wrong with psychology anyway? In: J.D. Cichetti, W.M. Grove (eds.) *Thinking clearly about psychology: matters of public interest, vol. I*, Minneapolis: University of Minnesota Press, pp. 3–39.
15. Gaffan, D. (2001). What is a memory system? Horel's *critique* revisited. *Behavioural Brain Research, 127*: 5–11.
Horel, J.A. (1978). The neuroanatomy of amnesia: a critique of the hippocampal memory hypothesis. *Brain, 101*: 403–445.
Horel, J.A. (1994). Some comments on the special cognitive functions claimed for the hippocampus. *Cortex, 30*: 269–280.
Vanderwolf, C.H. and Cain, D.P. (1994). The behavioral neurobiology of learning and memory: a conceptual reorientation. *Brain Research Reviews, 19*: 264–297.
16. A sampling of references to studies of the neural basis of attention:
a) Peripheral filtering of non-attended stimuli: Hernández-Peón, R., Scherrer, H., and Jouvet, M. (1956). Modification of electrical activity in cochlear nucleus during "attention" in unanesthetized cats. *Science, 123*: 331–332.
b) Thalamic intralaminar nuclei and brain stem reticular formation: Jasper, H.H. (1960). Unspecific thalamocortical relations. In: J. Field, H.W. Magoun and V.E. Hall (eds.) *Handbook of physiology, Section 1: Neurophysiology, volume 2*. Washington, D.C. American Physiological Society, pp. 1307–1321.
Lindsley, D.B. (1960). Attention, consciousness, sleep and wakefulness. In: J. Field, H.W. Magoun, and V.E. Hall (eds) *Handbook of physiology, Section 1: Neurophysiology, volume 3*. Washington, D.C. American Physiological Society, pp. 1553–1593.
c) The hippocampus: Bennett, T.L. (1975). The electrical activity of the hippocampus and processes of attention. In: R.L. Isaacson and K.H. Pribram (eds). *The hippocampus, volume 2: Neurophysiology and behavior*. New York: Plenum Press, pp. 71–99.
d) The cingulate cortex: Kaada, B.R. (1960). Cingulate, posterior orbital, anterior insular and temporal pole cortex. In: J. Field, H.W. Magoun, and V.E. Hall (eds). *Handbook of physiology, Section 1: Neurophysiology, volume 2*. Washington, D.C. American Physiological Society, pp. 1345–1372.

e) Parietal and frontal cortex: Colby, C.L., and Goldberg, M.E. (1999). Space and attention in parietal cortex. *Annual Review of Neuroscience, 22*: 319–349.

Kastner, S., and Ungerleider, L.G. (2000). Mechanisms of visual attention in the human cortex. *Annual Review of Neuroscience, 23*: 315–341.

Kolb, B. and Whishaw, I.Q. (2001). *An introduction to brain and behavior.* New York: Worth Publishers, (see pp. 537–539).

f) Cholinergic projections from the basal forebrain: McGaughy, J., Everitt, B.J., Robbins, T.W., and Sarter, M. (2000). The role of cortical cholinergic afferent projections in cognition: impact of new selective immunotoxins. *Behavioural Brain Research, 115*: 251–263.

g) Ascending noradrenergic projections: Aston-Jones, G., and Bloom, F.E. (1981). Activity of norepinephrine-containing locus coeruleus neurons in behaving rats anticipates fluctuation in the sleep-waking cycle. *Journal of Neuroscience, 1*: 876–886.

h) Hippocampal long-term potentiation: Shors, T.J., and Matzel, L.D. (1997). Long-term potentiation: What's learning got to do with it? *The Behavioral and Brain Sciences, 20*: 597–655.

Amygdala: Gloor, P. (1960). Amygdala. In J. Field, H.W. Magoun, and V.E. Hall (eds). *Handbook of physiology, Section 1: Neurophysiology, volume 2.* Washington, D.C. American Physiological Society, pp. 1395–1420.

j) Pyriform cortex: Freeman, W.J., and Skarda, C.A. (1985). Spatial EEG patterns, nonlinear dynamics and perception: the neo-Sherringtonian view. *Brain Research Reviews, 10*: 147–175.

k) Superior colliculus: Goldberg, M.E. and Wurtz, R.H. (1972). Activity of superior colliculus in behaving monkey. II. Effects of attention on neuronal responses. *Journal of Neurophysiology, 35*: 560–574.

17. Rosenfield, L.C. (1968). *From beast-machine to man-machine,* New York: Octagon Books, p. 54.

18. Pfluger, E. (1853). *Die sensorichen Functionen des Ruckenmarks der Wirbelthiere,* Berlin: August Hirschwald. According to this concept, a spinal flexion reflex elicited by a pinprick is associated with a spinal awareness of pain.

19. Griffin, D.R. (1984). *Animal thinking,* Cambridge, MA: Harvard University Press.
Margulis, L. and Sagan, D. (1995). *What is life?* New York: Simon and Schuster.

20. Romijn, H. (2002). Are virtual photons the elementary carriers of consciousness? *Journal of Consciousness Studies, 9*: 61–81.

21. Sebel, P.S., Bonke, B., Winogrod, E. (eds.) *Memory and awareness in anesthesia,* Englewood Cliffs, NJ: Prentice Hall, 1993.

22. Ostrum, A.E. (1994). The "locked-in" syndrome – comments from a survivor. *Brain Injury, 8*: 95–98.

23. Sperry, R.W. (1974). Lateral specialization in the surgically separated hemispheres. In: F.O. Schmitt and F.G. Worden (eds). *The neurosciences: Third study program.* Cambridge, MA: M.I.T. Press, pp. 5–19.

24. Popper, K.R., and Eccles, J.C. (1977). *The self and its brain,* Berlin: Springer-Verlag (see pp. 311–333).

25. Weiskrantz, L. (1986). *Blindsight: a case study and implications.* Oxford: Clarendon Press.

26. Goodale, M. and Milner, D. (2004). *Sight unseen.* Oxford, U.K.: Oxford University Press.

27. Kimura, D. (1993). *Neuromotor mechanisms in human communication.* New York: Oxford University Press.

II. An introduction to behavior for neuroscientists

Neuroscientists whose academic background is primarily in physical science, anatomy, biochemistry, genetics, physiology, pharmacology, etc., are likely to feel somewhat bewildered when they consider the function of the brain in general terms. It will seem obvious that the normal functioning of the brain is responsible for all aspects of human conduct and mental capacity but how can one make any progress in understanding this whole area? Psychology, considered as an academic field, is not taken seriously by many scientists: it appears to be widely regarded as consisting largely of equal parts of trivia and nonsense. The inevitable result for many scientists is an unquestioning acceptance of commonsense views of the mind and human behavior. However, when the origin of these commonsense views is examined, it becomes apparent that they are derived, not from any form of scientific investigation, but from the speculations of ancient Greek philosophers, especially Aristotle and Plato (see Chapter I, The mind and the explanation of behavior). This is not reassuring. Considering the success rate of the ancient philosophers in physics, chemistry, physiology, etc., why should we trust their judgment in the field of the mind and human behavior?

One of the great benefits of studying history, especially the history of science, is that we become aware that highly intelligent people in past centuries accepted beliefs that we now know to be completely false. This prompts the thought that some of the things we believe today will also be regarded as nonsense by our descendants. Is it possible that today's conventional opinions about the mind and human behavior will, at some point in the future, appear to have much the same validity and authority as is now granted to alchemy, astrology, and Ptolemaic astronomy?

Let us attempt to think through the problem of behavior and the mind very carefully. First of all, possession of the power of movement is one of the most striking characteristics of animals. Among the multicellular organisms, individual plants and fungi remain rooted in one spot throughout life. If local conditions became unfavorable, they must adapt as best they can, relying on genetic and physiological defences. Although these reactions are ordinarily very slow, it is most impressive that higher plants can coordinate the activities of a variety of different tissues hormonally by means of auxins, cytokinins, and gibberellins without having anything resembling animal nervous tissue.

In contrast to plants, animals (except a few sessile forms such as sponges or barnacles), when confronted with unfavourable conditions, can move away relatively quickly in the hope of finding something better. Since mere random motor activity may make things worse rather than better, there has evidently been a strong selection pressure favoring the development of sensory organs and nervous centers to guide and control motor activity.

It is conventional to refer to motor activity in a general sense by the term "behavior". This includes primarily posture and movement. Thus, holding the head up against gravity, sitting up, standing, walking, speaking, etc., are common components of human waking behavior; lying down with eyes closed and with a relaxation of postural tone are common aspects of sleep behavior. At times there have been attempts to distinguish between "behavior" and "physiological reactions" such as shivering or simple somatomotor reflexes. Such distinctions seem to me to be purely arbitrary and based on an implicit assumption that some motor patterns are the result of psychic or mental activity but others are not. It is simpler to assume that all motor activity is the result of physiological activities and to refer to the entire class of motoric and postural activities as "behavior." Whether autonomic activities should also be considered to be behavior is a matter of taste. Is blushing a behavior? What about piloerection or sweating in response to social stresses?

Systematic study of behavior developed in the late nineteenth and early twentieth centuries in three geographic regions: (a) the Sechenov-Pavlov school of reflexology in Russia; (b) the ethology-animal behavior school of Heinroth, Lorenz, and Tinbergen in Western Europe; and (c) the behaviorist school of Thorndike, Jennings, Watson and Skinner in America. In addition, studies of reflex activity, especially by Sherrington in England and Magnus in the Netherlands provided an essential foundation for our understanding of the physiological basis of simple behaviors.[1] Two essential assumptions underlay all of these varied endeavors: (1) motor activity should be recorded and observed in objective terms, avoiding all subjective psychological interpretations; and (2) all behavior is due to the physical and chemical activity of sense organs, neurons and muscles. Interpretations of behavior that depended on the activities of a non-material mind or psyche were ruled inadmissible.[2] It is widely assumed in this field that what requires explanation is behavior itself rather than some mental process that may be hypothesized to underlie behavior.

An important concept in modern studies in the science of animal behavior that coalesced out of the work of the pioneers in the field is that the varied behaviors displayed by an animal have evolved under the influence of natural selection. Therefore, even infrequent and seemingly trivial aspects of behavior are likely to have a real biological function and are well worth the attention

of serious investigators. For example, Niko Tinbergen devoted a not incon-
siderable research effort to determining why black-headed gulls carry empty
egg shells away from the nest shortly after the chicks have hatched.[3] This is a
behavior that may occupy no more than a few seconds per year. Nonetheless, it
has an adaptive role in the life of black-headed gulls and must have a definite
neural basis.

What this means for neuroscience is that all aspects of behavior must be
studied, including not only behaviors of obvious importance such as feeding
or reproductive behavior, but also behaviors whose contribution to adaptation
may not be immediately obvious. One can think of this as a three-stage process.
First, careful observation of spontaneous behavior is required to determine
what animals do in terms of the actual postures and movements that are
displayed. Second, controlling factors such as current stimulus input, levels
of nutrients, electrolytes, or hormones, body temperature and past experience
should be identified. Third, the role of different brain regions, different types of
central neurons and different neurotransmitters or intracellular signals should
be identified using various neuroanatomical, electrophysiological, neurochem-
ical, neuropharmacological, and brain imaging techniques. In all such work, it
is essential that a broad spectrum of behavior should be investigated, including
all aspects of feeding, reproductive behavior, social and parental behavior,
avoidance of natural dangers (including predators), body grooming, sleep, and
shelter-seeking behaviors (which play a major role in temperature regulation).
When dealing with human subjects, in particular, the "social behavior" category is a very large topic indeed, encompassing language, gestures and facial
expression.

Disentangling behavior from the psyche. Although it is today a common
belief that scientific progress is dependent largely or entirely on the development of new technologies, a modest degree of acquaintance with the history
of science reveals that possession of appropriate concepts and theories is of
even greater importance. It is quite possible to spend years making accurate
detailed measurements of things that are subsequently understood to be of no
consequence whatever. The medieval and early modern alchemists possessed
an impressive array of chemical techniques including solution, calcination,
sublimation, fusion, crystallization, distillation and fermentation but made only
slow and accidental advances in chemistry because their efforts were directed
towards the discovery of the philosopher's stone (to transmute base metals into
gold) and the elixir of life (to confer immortality).[4]

I think that advances in understanding the function of the brain have been
similarly impeded by continued adherence to an ancient and inappropriate set
of concepts. It is of great importance to make a clear distinction between behavior and hypothetical mental processes offered as explanations for behavior.
Thus, speaking is a behavior; the cognitive processes that may be invoked to

account for the speech are not behavior. Scratching one's head is also a behavior but a sensation of itchiness in the scalp is not. The essential distinction here is that "behavior" is a physical event that can be observed externally or detected by a recording device of some sort. Subjective states, by their very nature, cannot be detected by an external observer.

These distinctions are fundamental to any general approach to the function of the brain. If one thinks that the overall aim is to account for behavior then one must first make a catalogue of the behaviors a given species displays and then begin an analysis of central nervous control of those behaviors.

In contrast to this, a mentalistic approach suggests that the only behavior patterns that are worthy of serious study are those that can be assumed to be indicative of the activity of some mental process such as attention, cognition, emotion, or memory. It is assumed that we already have a good knowledge of the nature of these processes: therefore we can devise behavioral tests to measure them on an *a priori* basis. For example, such tests as delayed match to sample or delayed non-match to sample were widely adopted because they seemed to provide rather pure tests of memory which was conceived of as a mental process distinct from sensation, perception, attention, motivation and motor processes.[5] The difficulties and lack of real progress associated with this approach have been discussed in more detail elsewhere[6] (also see Chapter I).

The conventional theory of the brain as the organ of the psyche or mind offers us the comforting illusion that we already understand the big picture. We know how the brain/mind works because Plato, Aristotle and Descartes analyzed it for us long ago. If we abandon this, we become acutely aware of the enormity of our own ignorance. We must begin almost at the beginning, carefully analyzing brain activity in relation to behavior, tentatively feeling our way and building on our successes. My own conviction that this is the only possible way of making advances in the brain-behavior field is based, not merely on arguments of a semi-philosophical nature, but also on more than four decades of experience on the relations between behavior and the electrophysiological activity of the hippocampus, the neocortex, and the pyriform cortex. During the course of this work it became ever more apparent that brain field potential activity and the related unitary activity are not organized in terms of conventional psychological concepts but are, rather, closely related to various sensori-motor processes.[7] Mentalistic approaches to the brain-behavior field discourage the discovery of the relations between brain activity and sensori-motor processes because they: (1) encourage the belief that the details of behavior are trivial and unworthy of serious scientific study; and (2) encourage investigators to ask inappropriate questions. Neuroscientists who are interested in the overall function of the central nervous system should acquaint themselves with the study of behavior in both human and non-human animals and should learn to recognise the nature and present day influences of ancient philosophical theories concerning the psyche.

Notes

1. Short histories of the study of animal behavior have been provided by: Lorenz, K.Z. (1981). *The foundations of ethology*, New, York: Springer-Verlag, and by: Ratliff, F. (1962). Some interrelations among physics, physiology and psychology in the study of vision. In: S. Koch (ed.) *Psychology: A study of a science. Study II. Empirical substructure and relations with other sciences vol. 4: Biologically oriented fields: Their place in psychology and biological science.* New York: McGraw-Hill, 417–482. A collection of landmark papers in the history of animal behavior has been provided by: Houck, L.D., and Drickamer, L.C. (editors) *Foundations of animal behavior*, Chicago: University of Chicago Press, 1996. Useful summaries of classical reflex physiology and its relation to behavior include: Denny-Brown, D. (1939). *Selected writings of Sir Charles Sherrington*, Oxford, U.K.: Oxford University Press; Fukuda, T. (1984). *Statokinetic reflexes in equilibrium and movement*, Tokyo: University of Tokyo Press; and Fulton, J.F. (1949). *Physiology of the nervous system*, 3rded. New York: Oxford University Press.
   An excellent modern introduction to the behavior of the laboratory rat that is relevant to neuroscience is: Whishaw, I.Q., and Kolb, B. (editors) *The behavior of the laboratory rat: a handbook with tests*. Oxford: Oxford University Press, 2005. A very general discussion of recent developments in the Thorndike-Watson-Skinner approach to behavior has been provided by: Staddon, J. (2001). *The new behaviorism: mind, mechanism and society*, Philadelphia: Psychology Press.
2. It is interesting that Sherrington, who allowed no trace of mentalistic interpretations in his studies of reflexes, was nonetheless a dualist and believed that higher level perceptual and motor processes involved something beyond anatomy and physiology.
3. Tinbergen, N. (1972). *The animal in its world, vol. 1, Field studies*. London: George Allen and Unwin Ltd., pp. 250–294. It appears that gulls remove egg shells from the vicinity of the nest soon after hatching because the white interior of the empty shell attracts predators.
4. Holmyard, E.J. (1990). *Alchemy*. New York: Dover Publications (first published, 1957).
5. Vidyasagar, T.R. (1993). Assessment of brain electrical activity in relation to memory and complex behaviour, in *Methods in Neurosciences*, vol. 14, *Paradigms for the Study of Behavior* (Conn, P.M. ed.) Academic Press, San Diego, pp. 407–431.
   In delayed matching tests, three food wells (i.e. large holes drilled in a thick piece of plank or plastic) are placed just outside the bars of a cage containing a monkey. As the monkey watches, a food item is placed in the center well, covered by a distinctive item such as a beer can, and the animal is allowed to retrieve the food. An opaque screen is then lowered and the beer can (for example) is placed over one of the outside food wells and a novel object, such as an empty bottle, is placed over the other outside food well (either the right or the left one in random sequence). After a variable delay, the monkey is allowed to choose one of the objects. In the delayed match-to-sample version of this test the food item would be located under the beer can in this example while in the delayed non-match-to-sample version of the test the food would be located under the bottle. Thus, in everyday language, we can say that the monkey is required to remember the original demonstration item (many different items are used) and to make a choice based on that memory.

6. Vanderwolf, C.H. and Cain, D.P. (1994). The behavioral neurobiology of learning and memory: a conceptual reorientation. *Brain Research Reviews, 19*: 264-297. Also see: Vanderwolf, C.H., and Leung, L.-W.S. (1998). The relation of brain electrical activity to behavior. In: A.A. Boulton, G.B. Baker, and A.N. Bateson (eds.) *Neuromethods, vol. 32, In vivo neuromethods*, Totowa, New Jersey, pp. 325–357.
7. Vanderwolf, C.H. (2003). *An odyssey through the brain, behavior, and the mind.* Boston: Kluwer Academic Publishers.

III. Brain organization and behavior: The big picture

The behavioral functions of the central nervous system are usually discussed in terms of conventional psychological categories, processes or faculties which are assumed to be localized in different parts of the brain. Some parts of the cerebral hemispheres are said to provide the basis of sensation and perception while other parts are said to provide the basis for emotion, attention, memory, abstract thought, and voluntary control. This theoretical scheme is based on a psychological tradition originating with Aristotle and his predecessors. Since Aristotle believed that the psyche was associated particularly with the heart, it would truly be remarkable if the categories of the psyche which he discussed proved to be a valid description of the functional organizational of the brain. Rather, it seems probable that Aristotelian mentalistic concepts and their modern descendants have no more relation to the actual function of the brain than the Aristotelian chemical elements of fire, water, earth and air have to the subject matter of chemistry.

If the conventional mentalistic interpretation of cerebral function is truly invalid, we must approach the problem from a different direction. The main alternative approach to understanding higher level brain function seems to be to begin at the beginning by direct observation of the movements and postures (behavior) displayed by animals. Our task, then, is to discover how the brain generates all these movements and postures and how they are controlled by such factors as sensory inputs, hormonal conditions, and the effects of past experience.

If we attempt to gain a very general overview of how the brain generates behavior, it immediately becomes apparent that most behaviors involve the coordinated activity of the entire nervous system. The simplest approach to studying the behavioral capacities of different parts of the central nervous system involves surgical isolation of one part from the remainder (see Figure III.1). Although information of this type has been available for decades, its relevance to understanding the neural basis of behavior has not been widely appreciated. If the spinal cord, or a considerable part of it, is separated from the brain by a transverse cut, and an interval of time is allowed for recovery, various reflexes can be readily elicited by appropriate stimuli.[1] For example, if a noxious stimulus (such as a pinprick in a toe pad) is presented to the hind

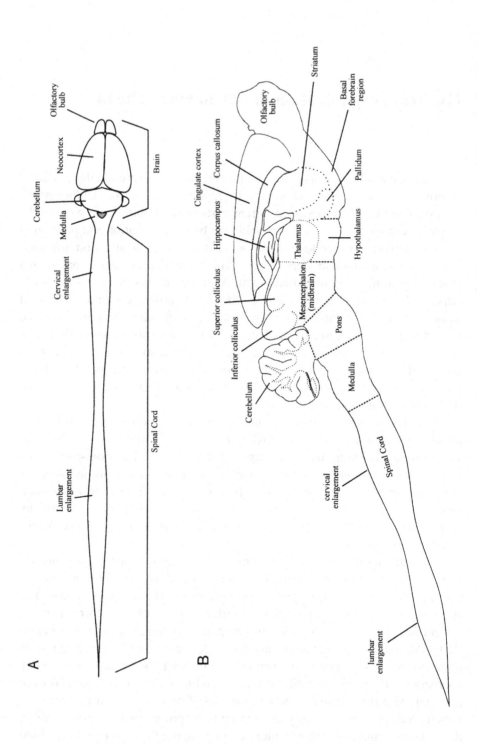

leg of a chronic spinal dog, the stimulated leg reacts by flexion at the hip, knee, and ankle joints (flexion reflex). At the same time the opposite hind leg displays extension at the hip, knee, and ankle joints (crossed extensor reflex). The overall pattern can be thought of as a spinal component of a defensive or protective behavioral reaction in which the injured limb is withdrawn while the opposite limb extends to bear the weight of the body. The scratch reflex, a component of the normal behavior of grooming the fur, can be readily elicited in a spinal dog by a light moving tactile stimulus (such as dragging the corner of an index card through the fur) which mimics the effects of a louse or flea crawling over the skin.

If a spinal dog is suspended vertically with the hind legs hanging free, alternating stepping movements occur in them (mark-time reflex). Reflex stepping movements, alternating in the two hind limbs, can also be elicited by placing the hind paws in contact with a moving treadmill. Pushing a finger-tip between the toe-pads and the plantar cushion of one hind paw in a spinal dog (thereby spreading the toes as would occur naturally when the paw is placed on the ground) elicits a strong extensor thrust reflex which would normally help to support the weight of the body and move it forward. These reflexes

Figure III.1. The central nervous system of the rat. *Top*: dorsal view of the brain and spinal cord. Drawn from a photograph in: Vanderwolf, C.H., and Cooley, R.K. (1990). *The sheep brain: A photographic series*, London, Ontario: A.J. Kirby Co. *Bottom*: A longitudinal section through the central nervous system has divided it into right and left parts (parasaggital plane, the cut is somewhat to one side of the midline). Major subdivisions are outlined by dotted lines. The medulla, pons, and midbrain together constitute the brain stem; the thalamus and hypothalamus together constitute the diencephalon; the neocortex, cingulate cortex, hippocampus and pyriform cortex (on the ventral surface of the brain, not shown here) together constitute the cerebral cortex; the cerebral cortex, striatum, pallidum, basal forebrain region and diencephalon are included in the forebrain. The striatum (which includes the caudate nucleus and the putamen) and the pallidum (also known as the globus pallidus) plus the substantia nigra (located in the ventral midbrain) are often referred to collectively as the basal ganglia. The forebrain and brainstem are connected to 12 pairs of cranial nerves which have various sensory and/or motor functions. The spinal cord is connected to 31 pairs of spinal nerves; each nerve is attached by a dorsal root containing sensory fibers and a ventral root containing mainly motor fibers which innervate the muscles. The cervical enlargement governs the functions of the forelimbs, the lumbar enlargement governs functions of the hindlimbs, and the narrow intermediate part governs functions of the thorax. A surgical transection through the thoracic or low cervical levels of the cord disconnects the lower part from the remainder of the central nervous system, permitting a study of its behavioral capacities in isolation (spinal animal). A similar transection dividing the midbrain from the forebrain permits study of the behavioral capacities of the spinal cord, brain stem and cerebellum in isolation (high decerebrate or midbrain animal).

probably function as components of locomotor behavior in a normal intact animal. Gentle manipulation of the genitalia readily elicits penile erection and a forward thrusting of the pelvic region in male spinal dogs and cats. Gentle mechanical stimulation of the clitoris and the walls of the vaginal orifice in spinal female dogs and cats elicits contractions of the uterus.

Similar reflex phenomena occur in humans who have had the misfortune of having the spinal cord severed by an injury (from a bullet wound, for example). Spinal humans display a flexion reflex and crossed extensor reflex. Penile erection is readily obtained in most chronic spinal men by gentle rubbing of the glans and frenulum of the penis. Ejaculation can also be elicited and there are said to be a number of cases of fatherhood in spinal men. Therefore, some of the basic reactions involved in defensive, locomotor, body grooming, and reproductive behavior are organized by neural circuits located in the spinal cord.[2]

When one considers that the mass of the spinal cord constitutes about 14 percent of the central nervous system (brain plus spinal cord) in a dog and only about 2 percent of the central nervous system in humans,[3] these observations seem remarkable. How can we interpret the fact that a great deal of behavior is based on spinal reflexes?

It is essential to consider what the isolated spinal cord cannot do as well as what it can do. For example, although reflex stepping and other locomotor reactions can be elicited in chronic spinal animals, true locomotion is not possible and there is no possibility of complex spontaneous behavior of any kind. The spinal cord cannot maintain an erect posture of the body (i.e., equilibrium cannot be maintained) and the tonus or sustained contractile power of the anti-gravity muscles may not be sufficient to prevent the body from sagging slowly to the ground.

Much more complex forms of behavior are possible if the brain stem and cerebellum are allowed to collaborate with the spinal cord. Thus, if the brain stem is transected along the line dividing the midbrain from the diencephalon (thalamus and hypothalamus, Figure III.1), the resulting high decerebrate animal displays a great variety of reflexive and spontaneous behaviors.[4] It can move the head about spontaneously or turn the head in response to a sound, and can walk about spontaneously. High decerebrate rats will also lick or nibble at objects which are brought in contact with the lips and teeth. They are also quite capable of swallowing. Nonetheless, despite having most of the reflexive bits of behavior necessary for feeding, they make no attempt to feed themselves. Similarly, other complex behavior patterns such as mating behavior, maternal care, or avoidance of dangerous situations are not present in an effective form. For example, high decerebrate animals will walk off the edge of a table without the slightest hesitation. We can conclude that neural circuits in the spinal cord, medulla, pons, cerebellum and midbrain in rats are capable of generating rather normal looking upright posture, head movement and locomotion but that nonetheless the normal behavior patterns of feeding, reproduction, avoidance of danger, etc., are grossly impaired owing to the absence of the forebrain.

What aspect of behavior is missing in these animals? How is it possible that an animal which, for example, is perfectly capable of biting and swallowing is, nonetheless, completely incapable of feeding itself? Consider the behavior of a food-deprived normal rat offered a piece of rat chow at a little distance. Olfactory and other sensory inputs activate the cerebral cortex which then, in turn activates brain stem and cerebellar circuits which, in their turn, activate spinal circuits generating locomotion and head movements which bring the rat's snout in contact with the food. The contact stimuli thus produced trigger mouth opening and biting reflexes which result in ingestion. We know that mouth opening and biting really are reflexive because surgical section of sensory nerves from the snout in an otherwise intact rat prevent biting even though the animal still approaches food or prey and places its snout in contact with it.[5]

A high decerebrate rat possesses reflexive biting and swallowing behavior but no longer possesses the cerebral mechanisms of the control of locomotion and head movement which normally guide its behavior. If locomotion and head movement no longer fulfil their normal function of placing the snout in contact with food, eating cannot occur.

The study of neuroanatomy has revealed the basic neural circuitry involved in these behaviors.[6] The spinal cord contains columns of large motor neurons which send out axons via the ventral roots to the large muscles of the shoulders, hips, and trunk (proximal musculature). The activity of these spinal motor neurons is controlled jointly by: (a) sensory inputs from the skin, muscles, tendons, joints and visceral structures; and (b) descending projections from the brain. The descending brain projections include: (a) vestibulospinal projections originating in the vestibular nuclei in the dorsolateral part of the medulla; (b) reticulospinal projections originating in the reticular formation located in the ventral and medial parts of the medulla and pons; and (c) tectospinal projections originating in the superior colliculus (tectum). A fourth pathway, containing fewer fibers but functionally related to the first three, arises from the interstitial nucleus of Cajal, located in the midbrain. These descending projections play an essential role in gross movements such as locomotion and head movement which depend on the activity of the proximal musculature. This is shown by several types of findings: (a) neurons in the reticular formation and other sites fire at high rates in correlation with gross movements; (b) surgical destruction of neurons in the reticular formation and vestibular nuclei abolish gross movements such as assuming an upright posture when placed on the back or side (righting) and locomotion; and (c) electrical stimulation of most sites in the ventromedial medulla and pons in freely moving animals gives rise to locomotor or other gross movements.

It appears to be the case then, that descending projections from the brain stem to the spinal cord are the primary means by which the brain is able to control behavior. However, the behavior which the brain stem, cerebellum, and spinal cord can generate when acting in isolation is dreadfully maladaptive and inadequate. An animal lacking its forebrain could not live long without extensive nursing care. Therefore, the forebrain must exert a decisive control over the activity of brain stem, cerebellar and spinal circuitry.

In neural terms, what all this means is that gross movements of the head, trunk, and limbs are controlled by a large number of specific spinal circuits which, when activated, produce isolated bits of behavior such as alternate stepping, scratching the body, forward thrusting of the pelvis, etc. Such circuits are often referred to as "central pattern generators." Pattern generators may be activated by a sensory input (producing a reflex) or by descending projections from the brain. Descending projections from the brainstem appear to activate combinations of pattern generators to produce a co-ordinated behavior such as walking forward, turning the head, rolling over, etc. The different brain stem circuits that produce such items of behavior are under the control of forebrain structures, especially the cerebral cortex. Thus, depending on such factors as current sensory input, hormonal or nutritional state, etc., the cerebral cortex will activate one or another brainstem circuit to produce turning right, turning left, standing motionless, etc.

If the cerebral cortex (neocortex, cingulate cortex, hippocampal formation, pyriform lobe) is surgically removed without extensive direct injury to the diencephalon (thalamus and hypothalamus) or the striatum, rats and other laboratory animals have a somewhat greater range of behavior than high decerebrate or midbrain animals. Such preparations display a grossly normal sleep-waking cycle, running about actively during the night and spending much of the day asleep, often in a curled-up nose-to-tail posture. Thus, the main features of sleep-waking behavior are organized subcortically. Further, unlike high decerebrate animals, decorticate animals are eventually able to feed themselves if food is easy to obtain. For example decorticate rats will eat a highly palatable food (lard mixed with brown sugar) if a large dollop is placed on a flat piece of metal but they are quite defeated if the lard-sugar mixture is placed in a flat dish with edges raised approximately one centimeter.[7] Similarly if adequate environmental support is provided, decorticate male rats may copulate successfully and decorticate female rats may succeed in raising a litter of young.[8] The behavioral state produced by extensive destruction of the cerebral cortex is widely known as "dementia".

If one searches the neuroscientific literature with the aim of discovering the neural basis of specific behaviors, one discovers a good deal of information but it is generally haphazard and unsystematic. Investigators preoccupied with the search for the neural basis of some hypothesized mental process have rarely done a thorough job of describing behavior. Studies on the role of the amygdala, a large cellular complex underlying part of the pyriform lobe, provide an example.

The amygdala are often said to be related to fear and anxiety. If this were really true one might expect that surgical removal of the amygdala would decrease fear or fearful behavior. However, although amygdalectomized monkeys are "fearless" in their tendency to approach and investigate objects avoided by normal monkeys, they are abnormally submissive and fearful in social situations with other monkeys.[9] Therefore, it cannot be said that "fear", in a general sense, is either increased or decreased by destruction of the amygdala. It is likely that there are a variety of different instinctive or learned sensorimotor behavior patterns which are altered or abolished by amygdalectomy. We cannot assume that the behavioral changes will conform to what might be expected on the basis of conventional psychological ideas. What is required is a far more detailed and comprehensive study of the actual behavior of brain damaged and normal animals than has usually been done in the past.

Role of the cerebral cortex in behavior. According to traditional ideas on the subject, the mammalian neocortex is subdivided into three broad types: (1) primary sensory cortex which receives an input from one or another of the sense organs such as the retina, the cochlea, or tactile receptors in the skin; (2) association cortex lying adjacent to two or more sensory cortices; and (3) motor cortex, a localized region responsible for cortical or voluntary control of movement. It is widely held that a sensory input produces a conscious sensation in the primary sensory cortex but that further elaborations, dependent on memory and associations with other stimuli, occur in association cortex. If a decision is made to make a movement, the association cortex can then activate the motor cortex.

Is this theoretical scheme, based on a psychological tradition originating with Aristotle and his predecessors, really a valid description of what happens in the brain?

First, since there is no known way of objectively determining when consciousness is present and when it is not, we really cannot tell whether or not conscious sensations occur in primary sensory cortex. Second, there is serious doubt that association cortex actually exists. According to I.T. Diamond, nearly all of the neocortex is divided into three great fields: an auditory field; a visual field; and a somesthetic field (Figure III.2). If one includes olfaction and visceral sensation as well, then the entire cerebral cortex is a target of sensory input of some kind. The significance of the "motor cortex" will be discussed in more detail below.

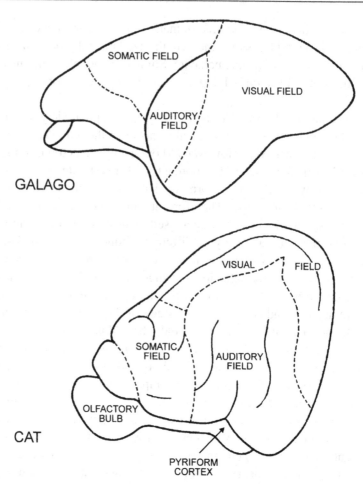

Figure III.2. Neocortical sensory fields in the primate and the cat. After Diamond, I.T. (1985). A history of the study of the cortex: changes in the concept of the sensory pathway, In: Kimble, G.A., and Schlesinger, K. (editors) _Topics in the history of psychology._ Hillsdale, New Jersey: Lawrence Erlbaum Associates, Inc., pp. 305–387. The entire neocortex is divided into an auditory field, a somatosensory field and a visual field. The orbitofrontal area (on the left, above the olfactory bulb), left unlabelled by Diamond, contains an olfactory field. Olfactory cortex also includes the entire pyriform lobe at the base of the brain plus the hippocampal formation (shown in Figure III.1).

Since the traditional scheme of the functional organization of the cerebral cortex is clearly not in agreement with the facts, some other scheme is required. Where can one begin?

Comparative anatomical and electrophysiological studies have shown that the cerebral cortex in primitive vertebrate brains is dominated by olfactory inputs and distributes its outputs to various brain stem structures. In reptiles, for example, although there are many projections from the cerebral cortex to the brain stem, there appear to be no direct pathways from any part of the cerebral cortex to the spinal cord.[10] This anatomical arrangement permits the cerebral cortex to control behavior pattern generators in the brainstem but does not allow direct cerebral control of circuits in the spinal cord. Thus, when a turtle or a lizard encounters the odor of food, carrion for example, olfactory inputs will activate cortical neurons. Descending cortical efferent fibers, directly or indirectly, then activate brain stem pattern generators for four-legged walking which, in turn, activate segmental spinal pattern generators to produce the co-ordinated locomotion needed to bring the reptile to the food. The overall course and pattern of locomotion will, naturally, be guided not only by olfaction but also by visual, auditory and tactile stimuli encountered along the route.

In mammals, unlike reptiles, there has been an extensive development of a new neural structure, the neocortex, which has direct projections to the spinal cord (corticospinal pathway). This suggests that the mammalian cerebral cortex can control behavior via two systems of descending projections. An ancient system, probably present in all vertebrates, is based on direct or indirect cortical projections to the brainstem. Thus, the hippocampal formation, a prominent and distinctive part of the mammalian cerebral cortex, has multiple projections to the diencephalon and the ventral midbrain (tegmentum).[11] The pyriform cortex, the primary olfactory cortex, also has prominent efferent projections to the brainstem. Similarly the neocortex has prominent corticoreticular projections to the brainstem reticular formation, corticotectal projections to the tectum (superior colliculus) and corticostriatal projections to the striatum (caudate nucleus and putamen). Further descending fibers from the striatum can influence tectospinal fibers via a synaptic relay in the substantia nigra in the midbrain. The corticopontine projection, a massive system of fibers that includes efferents from all neocortical regions, provides a route by which the neocortex can influence not only pontine reticulospinal projections but also the pontocerebellar projections to the cerebellum. These various projections to the brain stem and cerebellum permit cortical control of pattern generators for gross motor activities such as locomotion, postural adjustments and orienting movements of the head.

In addition to this phylogenetically ancient cortical control of gross movement, mammals have evolved an additional system for the performance of discrete movements. This system presumably evolved to allow ancestral mammals to use the limbs, especially the forelimbs, for non-locomotor uses which require individual cortical control of a single limb or even of isolated digits. Thus, a cat can use a forepaw to strike at prey (or bat a ball) while a rodent or a primate can manipulate small pieces of food, a behavior that is greatly facilitated by separate control of individual digits. In contrast, reptiles such as the various turtles, lizards, and crocodilians appear to make very little or no use of one limb or one digit in isolation to manipulate objects.

The anatomical basis of neocortical control of discrete movement consists of descending projections to ventral horn cells (motor neurons) in the dorsolateral part of the ventral horn of the spinal cord. These cells activate muscles controlling the distal parts of the limbs. The descending fibers include: (a) corticospinal fibers originating in the neocortex; and (b) rubrospinal fibers originating in the red nucleus in the midbrain. Corticorubral projections permit cortical control of the rubrospinal pathway. Section of corticospinal fibers in rhesus monkeys has very little effect on gross motor patterns such as climbing, walking, running, jumping, turning the head or changing posture. However, in the immediate post-operative period there is a marked inability to move a limb in isolation, making it difficult or impossible to reach out, grasp, and lift a piece of food while sitting immobile. Although there is some recovery of reaching, grasping, and lifting with the passage of time, independent movements of the digits never recover. Thus, monkeys sustaining an extensive interruption of corticospinal fibers can make a grasping movement by flexing all the fingers together but cannot extend the index finger and thumb to make a precision grip (for example, to pick up a raisin or peanut) while keeping the other fingers flexed. It appears that the corticospinal system (probably assisted by the rubrospinal system) is involved in discrete cortical control of restricted parts of the body.

Experiments by Leyton and Sherrington early in the twentieth century showed that in deeply anesthetized chimpanzees, gorillas, and orang-utans, movements could be elicited by localized electrical stimulation only in restricted regions of the neocortex. Movements of the limbs and body could be elicited from the precentral gyrus in the frontal lobe but movements of the eyes or eye-lids could also be elicited from a more rostral region in the frontal lobe and from a part of the occipital lobe. This experiment, and others using similar methods, were very influential in establishing the concept that there is a localized motor cortex in the frontal region of the neocortex. Later experimenters who used unanesthetized or more lightly anesthetized animals have found that movements of one kind or another can be elicited by localized electrical stimulation of virtually all regions of the neocortex.[12] The differing results are probably due to the use or non-use of anesthetic drugs which have the effect of depressing synaptic transmission. The precentral gyrus, which has strong access to spinal motoneurons over a pathway involving only one or two synapses would be relatively unaffected by anesthetic but other cortical regions, the occipital lobe, for example, which has access to spinal motor neurons only over multi-synaptic output pathways, involving the striatum (corticostriate fibers), the tectum (corticotectal fibers), or the reticular formation (corticoreticular fibers) would be more strongly affected by anesthetics. It appears, therefore that "motor" functions are not restricted to a small precentral cortical region but are instead located throughout the neocortex. It is helpful to think about this in anatomical terms. The neocortex consists of several distinctive cell layers. The more superficial layers which contain large numbers of small pyramidal cells or granule cells receive the main sensory inputs from the thalamus while the deep layers contain larger pyramidal cells that send axons outside the neocortex. The large pyramidal cells of layer V, in particular, appear to be responsible for all neocortical projections to subcortical structures other than the thalamus. All neocortical areas, then, have a sensory or input zone, interneurons, and a set of efferent or motor cells. The "visual" neocortical area is really a "visuomotor" area, the "auditory" neocortical area is really an "audiomotor" area, and so on. Similarly, in the hippocampal formation, the dentate granule cells are the main target of the olfactory inputs to this structure while the large pyramidal cells of Ammon's horn provide an efferent output. Therefore, the entire cerebral cortex can be thought of as having both sensory and motor functions but the traditional "sensori-motor" areas of the neocortex have a particular role in the performance of discrete movements.

The differing roles of the different neocortical areas are revealed quite clearly when they are removed. Experimental removal of the sensori-motor cortex in the rat has very little effect on gross movements such as walking or turning the head but produces severe impairments in the ability to reach for food, pick it up and manipulate it with one paw. On the other hand large neocortical lesions in *any* part of the neocortex produce clear deficits in directed locomotion, as is required for example in such behaviors as hoarding food, running through a maze or following and mounting a sexual partner. Although behavioral deficits of this type are often given complex psychological interpretations, they can be viewed more simply as indicative of a defect in the cortical control of locomotion which is normally exerted via sensory control of corticostriate, corticotectal, or corticoreticular projections. For example, cats in which the auditory cortex has been surgically removed may be quite capable of turning the head accurately toward a sound (probably a reflexive behavior involving the midbrain tectum) but cannot walk accurately toward a sound.[13]

From a functional point of view, it is interesting that the major source of corticospinal projections lies in a region of cortex that receives somesthetic and proprioceptive inputs. There is a very close relation between sensory input and motor output. A cortical motor neuron which produces adduction of the thumb when it is electrically stimulated receives an input from skin on the medial surface of the thumb; a different cortical neuron that produces flexion of the thumb when it is stimulated receives an input from skin on the palmar surface of the thumb.[14] Thus, a cortical motor neuron controlling a particular movement receives input from a skin area which is likely to receive stimulation as a result of that movement. It is likely that this arrangement evolved because of the advantages of close tactile or proprioceptive control of delicate manipulatory movements.

Other sense modalities such as vision, audition, or olfaction are not so closely related to discrete movements probably because these sense modalities are usually involved in the control of gross motor activities such as approaching food, fleeing a predator etc. However, mechanisms for close visual control of discrete motor activity have evolved in some animals, especially primates, but also in cats which use a forelimb to strike at visually located prey. Tract-cutting experiments have indicated that this kind of visuomotor control is exerted by descending pathways from visual cortex to some subcortical structure which then activates the classical motor areas via a second ascending pathway.[15] The details of these pathways remain to be discovered.

What does the sensory cortex do? The conventional answer to this question is that sensory cortex constructs an internal representation of the outside world. There is, however, a logical difficulty in this. If the eye, for example, forms an image of the outside world and the visual cortex constructs another image based on information provided by the eye, what is it that looks at the cortical image? It is evident that the conventional theory is based on an implicit assumption of a mind or psyche, distinct from the brain, which can view representations of the outside world that form in the sensory cortices.[16] This is the theory of the imprisoned knower whose only contact with the outside world is provided by a bank of television screens. Is there any conceivable alternative to this theory?

Sensory neocortex receives two general types of input from subcortical structures. The thalamus provides a massive, finely differentiated input which permits cortical cells to respond selectively to detailed features of sensory stimuli. Visual cortical cells may respond selectively to lines or edges or more complex stimuli such as a human or animal face. Cells in auditory cortex may respond selectively to pure tones of specific frequencies or to vocalizations of conspecifics. In addition to these specific inputs, the entire neocortex and hippocampal formation receive a sparser but very widespread input from cells in the basal forebrain region and from cells in the brainstem which make use of the neurotransmitters acetylcholine and serotonin, respectively. If either the specific or widespread classes of input is blocked, the cortex cannot function and the experimental animal becomes grossly demented, behaving in much the same way as animals in which the entire cerebral cortex has been surgically removed.[17]

Thus, both large thalamic lesions and selective blockade of choliner-gic (acetylcholine-dependent) plus serotonergic neurotransmission produces a state of general dementia. It is likely that the cholinergic and serotonergic projections fulfil a general regulatory role with respect to cortical neural activity while the thalamic projections provide a much more detailed and specific input.

How, in general terms, does the whole thing work? There is reason to think that the entity or process known as "perception" is fundamentally an activation of motor programs. Someone looking at a complex figure in which a smaller figure is concealed, may not detect the hidden figure at first. When the figure is finally identified, the observer is immediately capable of doing a number of things that were impossible a moment before. The hidden figure can be pointed out, described verbally, or illustrated in a sketch.[18] As J.J. Gibson[19] pointed out, perception of something is knowing what one can do with it or what one can do with respect to it. Gibson invented the term "affordance" to describe the situation. Thus, water affords drinking, swimming, or boating, a small rock affords throwing, hammering, weighing down papers, etc. A thing is perceived when motor programs relevant to it have been activated. One can imagine that viewing a complex visual scene sets up patterns of activity in visual cortex which have an output to such structures as the striatum, the thalamus, the tectum and the reticular formation. This normally results in behavior which is adapted to the perceived environment but the details of how it is accomplished are, as yet, almost completely unknown.

Notes

1. Sherrington, C.S. (1906). *The integrative action of the nervous system*, New Haven: Yale University Press.
   Forssberg, H. (1979). On the integrative motor functions in the cat's spinal cord. *Acta Physiologica Scandinavica* (Supplementum 474), 1–56.
2. Fulton, J.F. (1949). *Physiology of the nervous system*, 3rd edition, New York: Oxford University Press.
3. Krompecher, S. and Lipak, J. (1966). A simple method for determining cerebralization brain weight, and intelligence. *Journal of Comparative Neurology*, *127*: 113–120.
4. Woods, J.W. (1964). Behavior of chronic decerebrate rats. *Journal of Neurophysiology, 27*: 635–644.
   Lovick, T.A. (1972). The behavioural repertoire of precollicular decerebrate rats. *Journal of Physiology (London), 226*: 4–6p.
5. Gregoire, S.E. and Smith, D.E. (1975). Mouse-killing in the rat: Effects of sensory deficits on attack behaviour and stereotyped biting. *Animal Behaviour, 23*: 186–191.
6. Kuypers, H.G.J.M. (1982). A new look at the organization of the motor system. In: H.G.J.M. Kuypers and G.F. Martin (eds). Anatomy of descending pathways to the spinal cord. *Progress in Brain Research, 57*: 381–403.
7. Personal observations.
8. Whishaw, I.Q. (1990). The decorticate rat. In: Kolb, B., and Tees, R.C. (eds) *The cerebral cortex of the rat*, Cambridge, Massachusetts: The MIT Press, pp. 239–267.
9. Horel, J.A., Keating, E.G., and Misantone, L.J. (1975). Partial Klüver-Bucy syndrome produced by destroying temporal neocortex or amygdala, *Brain Research, 94*: 347–359.
10. Ten Donkelaar, H.J. (1982). Organization of descending pathways to the spinal cord in amphibians and reptiles. In: H.G.J.M. Kuypers and G.F. Martin (eds) *Anatomy of descending pathways to the spinal cord. Progress in Brain Research, 57*: 25–67.

Belekhova, M.G. (1979). Neurophysiology of the forebrain. In: C. Gans, R.G. Northcutt, and P. Ulinski (eds.) *Biology of the reptilia, vol. 10, Neurology B*, London: Academic Press, pp. 287–359.

11. Vanderwolf, C.H. (2001). The hippocampus as an olfacto-motor mechanism: were the classical anatomists right after all? *Behavioural Brain Research, 127*: 25–47.

12. Lilly, J.C. (1958). Correlations between neurophysiological activity in the cortex and short-term behaviour in the monkey. In: Harlow, H.F., and Woolsey, C.N., *Biological and biochemical bases of behaviour*, Madison: The University of Wisconsin Press, pp. 83–100.
    Neafsey, E.J. (1990). The complete ratunculus: output organization of layer V of the cerebral cortex. In: Kolb, B., and Tees, R.C. *The cerebral cortex of the rat*, Cambridge, Massachusetts: The MIT Press, pp. 197–212.

13. Heffner, H.H., and Masterton, R.B. (1975). Contributions of auditory cortex to sound localization in the monkey. *Journal of Neurophysiology, 38*: 1340–1358.

14. Rosen, I., and Asanuma, H. (1972). Peripheral afferent inputs to the forelimb area of the monkey motor cortex: Input-output relations. *Experimental Brain Research, 14*: 257-273.

15. Myers, R.E., Sperry, R.W., and McCurdy, N.M. (1962). Neural mechanisms of visual guidance of limb movement. *Archives of Neurology, 7*: 195–202.
    Penfield, W. (1954). Mechanisms of voluntary movement. *Brain, 77*: 1–17.

16. Stent, G.S. (1975). Limits to the scientific understanding of man: human sciences face an impasse since their central concept of the self is transcendental. *Science, 187*: 1052–1057.

17. Vanderwolf, C.H. (2003). *An odyssey through the brain, behavior, and the mind*. Boston: Kluwer Academic Publishers.

18. Sperry, R.W. (1952). Neurology and the mind-brain problem. *American Scientist, 40*: 291–312.

19. Gibson, J.J. (1979). *The ecological approach to visual perception*, Boston: Houghton Mifflin Company.

IV. Human origins and adaptations

Before one can begin to think seriously about the human brain and behavior in a general sense one must consider some fundamental questions. What are the basic characteristics of human behavior and how does human behavior compare with the behavior of other animals? An intelligent alien from a remote galaxy, observing modern humans for the first time, might well conclude that these creatures are, in the main, adapted for a crowded social life in an environment consisting largely of concrete, asphalt, glass and steel and that they feed largely on manufactured foods provided in plastic containers. More prolonged investigation, however would reveal that the urban environment now inhabited by many people is a very recent development and that throughout more than ninety-nine percent of its history humankind lived in small social groups of no more than a few dozen individuals subsisting on plant and animal products obtained in their natural state. Consequently, the natural condition for human beings is life as it was during the long paleolithic era prior to the development of agriculture. This must be the condition to which human behavior and the function of the human body are most closely adapted.

A rational understanding of the history of humankind has been achieved only in the last century or so. Traditional Christianity teaches that all of humanity descended from an original pair, Adam and Eve, specially created by God only a few thousand years ago. In opposition to this, Charles Darwin proposed in 1871 that mankind, over an immense period, descended from ape-like creatures and that our closest living relatives are the chimpanzee and gorilla. Since chimpanzees and gorillas are native to Africa, this suggested to Darwin that the human species probably originated there.[1] Although there was little direct evidence to support this idea in 1871, the subsequent discovery of numerous fossilized remains of ancient humans and of human-like creatures has strongly confirmed Darwin's hypotheses.[2]

In 1925 Raymond Dart, a teacher of anatomy, described the fossilized skull and lower jaw of a child, with characteristics intermediate to those of apes and humans, which had been found in a limestone mine near Taung in South Africa. This, and similar fossils subsequently discovered elsewhere in South and East Africa (Ethiopia, Kenya, Tanzania), have been shown to range from about 2.5–5.0 million years old. The creatures whose remains had been preserved in this way have become known as austrolopithecines after the name suggested by Dart for the Taung child (*Australopithecus africanus*). Most australopithecines were rather small creatures, probably about 100–150 cm in height and 30–60 kg in weight. The size of the canines and the prognathism, or forward projection of the lower face to form a muzzle, were reduced in comparison to apes or monkeys. The brain volume was about 400–519 cc which is comparable to present-day gorillas (about 425 cc) and chimpanzees (about 320–336 cc). Modern human brains usually range from 1300–1460 cc.[3] Despite the rather small brain of the australopithecines, there is clear evidence from the structure of the skull, the forelimbs, the pelvis, and the foot that they walked erect and bipedally, much as we do. This anatomical conjecture was dramatically confirmed by the discovery of a series of footprints made by three hominids walking in a fresh ash fall from a nearby volcano (Sadiman, near Laetoli, in Tanzania) that occurred between 3.49 and 3.76 million years ago.[4] The footprints closely resemble those one can see in the sand of any present-day beach where modern humans congregate.[5] This is an important finding because it demonstrates that obligatory bipedalism, one of the distinctive characteristics of humankind, developed long before the evolution of a large brain, another distinctive human characteristic.

Paleoanthropologists have suggested the former existence of a variety of different species of australopithecines including a rather fine-boned or gracile type (*A. afarensis, A. africanus, A. anamensis*) and a heavier "robust' type (*A. boisei, A. robustus, Zinjanthropus boisei, Paranthropus sp., and A. aethiopicus*). It appears to be true that different species of australopithecines with different habits and different diets were alive at the same time in the same general region of Africa.

In more recent geological strata, deposited about 2 million years ago, australopithecine fossils are replaced by hominid fossils presenting evidence of a larger brain, smaller teeth and a hand that must have been capable of a good thumb-against-finger tip grip (precision grip). An early group of fossils of this type, referred to as *Homo habilis* or *Homo rudolfensis* was apparently replaced by taller, bigger brained creatures known as *Homo ergaster, Homo erectus* and *Homo sapiens*, the latter being similar to present day humans. Some paleoanthropologists consider that *Homo habilis* and *Homo rudolfensis* are more appropriately referred to as *Australopithecus habilis* and that later species of *Homo* should all be referred to as *Homo sapiens*. The rather confused state of the terminology in this field will, no doubt, be resolved when a much larger number of hominid fossils have been discovered. For the moment, it appears that we have an ancient australopithecine group which was gradually replaced by a *Homo* group which was taller, heavier, had smaller teeth and jaws, relatively longer legs and shorter arms, a flatter face, and a much larger brain. *Homo* (or *Australopithecus*) *habilis* had a brain volume of about 640 cc. while

Homo erectus (also known as early *Homo sapiens*) had a brain volume of 895–930 cc.[3] This is at least twice the size of the brain in any living non-human primate but is still substantially smaller than the brain in modern humans. Humans with a fully modern anatomy appeared in roughly the last 100,000–250,000 years.

Hominid fossils begin to appear outside Africa, as far a field as Indonesia, as early as 1.5–2.0 million years ago. Evidence of the presence of humans does not appear in Europe north of the Pyrenees earlier than about 700,000 years ago. Neanderthal man, a unique type with a very robust skeleton, prominent brow-ridges, a noticeable occipital protuberance or "bun", and a receding forehead but with a brain fully as large as modern humans, appears to have been indigenous to Europe and the Middle East. Australia was colonized as early as 115,000 years ago but humans seem to have arrived in the Americas only 10–20,000 years ago.

One of the outstanding characteristics of humans as compared to other animals is a strong tendency to make and use tools. It is true that a variety of animals use simple tools but never to the extent that humans do. Wild chimpanzees, for example, will break nuts by placing them on a stone and hammering them with another stone. They also strip the leaves from a stem of grass or a twig to make a tool which can be inserted into ant or termite nests. When the insects attack this "fishing pole", the chimpanzee pulls it out and licks them off, or having stripped them off with one hand, licks them up from the hand. Chimpanzees and other apes have never been observed to make tools from stone or other hard materials such as bone, but early mankind did so regularly.[6]

When human ancestors first began to make tools from organic materials such as sticks or grass is unknown since such artifacts are rarely preserved as fossils. The earliest known stone tools have been excavated in Olduvai Gorge in Tanzania in geological strata as much as 2.4 million years old. The artisans of this Oldowan Industry, as it has become known, made edged tools by flaking pieces from pebbles or small stones. The edge was produced by hammering flakes from both sides of a stone so that the two exposed faces met at an acute angle. Rocks broken by natural processes generally do not have this bifacial appearance. Furthermore, microscopic study of the sharp edges of Oldowan tools reveal a pattern of wear that can be duplicated in modern experimentally-produced stone tools by such activities as scraping wood or skin, cutting meat and so forth. Different types of use produce recognizably different types of wear.

Gradually the making of stone tools became more systematic. A type referred to as Acheulean tools comprise large numbers of bifacially flaked pear-shaped objects that have a sharp point. These are generally referred to as "hand-axes", under the supposition that they were held in the hand, without an attached handle, and used to chop or cut a variety of materials. Although they were first discovered at Saint-Acheul in France in 1854, their earliest known appearance in the geological record is in Ethiopia about 1.4 million years ago. They continued to be used without significant change for a million years or more. The appearance of Acheulian hand-axes was associated (very roughly speaking) with other technological advances. Cutting tools began to be resharpened when they had become dull by use. Specialized tools of various types began to appear. Evidence that fire was used deliberately by ancient man appears as early as 1.6 million years ago in Koobi Fora in Kenya.

The human use of tools is often related to the evolutionary development of the human hand. In the words of two eminent authorities on this topic, J.R. Napier and P.H. Napier, "Through natural selection, the opposition of the thumb prompted the adaptation of the upright posture and bipedal walking, tool-using and tool-making which, in turn, led to an enlargement of the brain".[7] Is this likely to be true? The question is somewhat more complex than it first appears. The development of a peripheral anatomy that permits good opposition of the thumb to the fingers is of little consequence unless there is a prior or concomitant development of brain circuits permitting individual control of the thumb and fingers in a variety of different patterns. Which is the more important and which is likely to have evolved first, the brain circuits or the peripheral anatomy of the limb?

An answer to this question is suggested by the clinical syndromes of phocomelia and amelia which became relatively common for a time following the introduction of the drug thalidomide nearly 50 years ago. Thalidomide is a hypnotic and sedative drug with an extremely wide safety margin (the difference between a hypnotic dose and a lethal dose) and very few side effects in human adults. It was released for unrestricted use in West Germany in 1958 and in other countries soon afterwards. This precipitated an individual, medical, and social disaster. Many women who took thalidomide during pregnancy subsequently gave birth to children with no limbs at all (amelia) or deformed limbs somewhat resembling seal flippers (phocomelia). Cases of particular interest to students of human evolution are those in which the upper limbs are grossly deformed or absent while the lower limbs are normal. Other body structures are generally normal. Phocomelic children with no thumbs or index fingers at all, and the remaining digits weak, deformed, or partially fused together, could, nonetheless, use their upper limbs to eat and write and often did well in school. Children with no upper limbs at all learned to use their feet to feed themselves using cups, spoons, etc., to button or unbutton clothing, remove clothing etc. It is apparent that in such cases "functional disability cannot be accurately deduced from knowledge of the structural defect".[8]

The thalidomide children demonstrate that a modern human central nervous system can generate a great deal of skilled tool use even when it is housed in a grossly defective body. If a chimera were created with a chimpanzee body and a human nervous system there can be little doubt that its behavior would be entirely human. These observations are consistent with the hypothesis that effective tool use and the brain circuits underlying tool use evolved long before the appearance of the modern human hand. Natural selection favoring individuals who were particularly adept in the manufacture and use of tools would then slowly transform an ape-like hand into a human hand.

Most living animals make little or no use of tools. Why was early man an exception? A possible answer to this question is suggested by two evolutionary principles with wide applicability. First, the diet of an animal, the way it makes its living, has an immense impact on its anatomy, physiology and behavior. Thus, the obvious differences between a deer and a wolf in terms of teeth, bones, guts, etc., as well as behavior are, to a great extent, explicable in terms of the diet to which these animals are adapted. Second, natural selection can usually modify behavior more rapidly than gross anatomical structures. One example of this is provided by the behavior of the woodpecker-finch on the Galapagos Islands west of Ecuador. There are no true woodpeckers on the Galapagos Islands. However, the woodpecker-finch, a small sparrow-like bird, lacks the unique anatomical adaptations of woodpeckers, including a long tongue capable of removing insects from deep narrow cavities in wood or bark. Consequently, it has acquired the behavior of rooting insects out of holes by means of a cactus thorn or twig held in the beak.[9] Similarly, the shrikes, song birds which have secondarily adopted a predatory and carnivorous life style, have had to make do with a peripheral anatomy that is not well adapted to the life of a bird of prey. Hawks, eagles and owls have powerful clawed feet with which they can seize and kill other animals by driving the claws into the body. The dead prey can then be dismembered by holding it in the feet and tearing off pieces with the bill. Shrikes, having only the relatively weak feet of a typical songbird, kill their prey (insects, small reptiles, birds, and rodents) by pecking at the head. The body of the prey is then pulled apart with the bill after impaling it on a thorny tree or a barbed wire fence.[10] In these cases, tool-using behavior seems to have evolved to compensate for a lack of appropriate peripheral anatomical structures. It is possible that in humans, as in woodpecker-finches and shrikes, tool use was developed in association with a change in the diet. Thus, in place of the teeth and claws of typical predators, humans evolved the brain circuits necessary to produce and use a variety of tools for the killing, dismemberment, and transport of prey animals.

Primates, for the most part, are herbivorous animals. Some, like the gorilla, subsist primarily on leaves, shoots, and stems. Chimpanzees eat mainly fruit, supplemented by leaves and other plant foods. Although chimpanzees are active hunters (see below), meat constitutes less than 10 percent of their diet. Humans in traditional hunter-gatherer societies, as well as in modern affluent societies, typically eat far more meat than this. Animal meat and fat are preferred foods in virtually all human societies.[11] Even when the diet is largely vegetarian, meat is likely to be eaten on festive occasions such as weddings.

In the most extreme case of human carnivory, the traditional diet of the Inuit or Eskimo people consisted of nothing other than animal tissue and water. The effects of such a diet become known to science as a result of the observations of Vilhjalmur Stefansson, a Canadian arctic explorer in the early twentieth century.[12] Stefansson lived for some time, in the period 1908–1912, with the Copper Eskimos, a people living north of Great Bear Lake in the North-West Territories who had never seen a white man before he arrived. The Copper Eskimo ate fish and killed caribou, feeding their dogs on the guts, liver, heart and tenderloin, and reserving most of the fat, marrow, and other muscle tissue for themselves. They were well aware that eating nothing but lean meat produces an illness characterized by weakness, head-ache and diarrhea. This condition, well-known in the Canadian North, was popularly called "rabbit starvation" because it inevitably put in an appearance if people were forced to live on a diet of snow-shoe hares (*Lepus americanus*) which have extremely little body fat in the winter. Inuit living entirely on a diet of lean meat and fat maintain a state of vigorous health with no deficiency symptoms of any kind and no evidence of unusual cardiovascular disease. It was at first suspected that the Inuit had some sort of special genetic adaptation to a totally carnivorous diet but this was disproved in an experiment in which Stefansson and Karsten Andersen, another arctic explorer, spent a year living under medical supervision on a diet of lean meat (left rare according to Inuit traditional practice which avoids destruction of vitamin C and other nutrients) and fat plus water, black coffee and tea (without milk, cream, or sugar). Both men remained in good health except that Andersen had a bout of pneumonia from which he recovered successfully. Prior to this test, Stefansson had lived in good health for a total of about 9 years on a diet of fat meat and water.

It is not widely understood today that a diet consisting of animal tissue and water provides all the essential nutrients required by humans. The content of the traditional Eskimo daily diet has been estimated as: carbohydrates, 10 gms; fat, 185 gms; and protein, 200 gms. In energetic terms, fat provides about 66 percent of the caloric value of such a diet while carbohydrates provide only 2 percent.

To most modern urbanized people, a diet of fat meat and water seems rather odd but throughout much of human prehistory prior to the development of agriculture, it would have been common in the temperate, boreal, and arctic regions of the world simply because there is very little else to eat for much of the year. Certainly, in early Canada both the aboriginal people and the European fur traders lived almost entirely on fish and game, either fresh or in the form of pemmican (the flesh of deer or bison cut in thin strips, dried in the sun, pulverized, thoroughly mixed with fat and stored in a bag made from the dried skin of the slaughtered animal).[13]

It is noteworthy that the Pleistocene epoch (about 2,500,000 to 10,000 years ago) was characterized by a great abundance of large mammals and a small widely scattered human population. In contrast to present day hunter-gatherers, who have in many cases been driven into rather marginal environments, early humans lived in a hunter's paradise. Animal tissue, the preferred food of humans, would have been readily available. There is reason to suspect that the invention of tool-assisted predatory behavior by humans was overwhelmingly successful. There appears to be a correlation between the appearance of humans in various parts of the world and the extinction of many prey species during the late Pleistocene.[14] Numerous species of large marsupials had become extinct in Australia by about 30,000 years ago, but in North America, large land mammals (such as mammoths, ground sloths, native horses and camels, and giant beavers) became extinct only 8–12,000 years ago. In New Zealand, large flightless birds such as the moa became extinct only in the last 1,000 years or so. In each case, these extinctions coincide very roughly with the arrival of *Homo sapiens* in the region in question. One may wonder whether the extensive modern human consumption of vegetable products and the development of agriculture about 10,000 years ago may have occurred in response to increasingly poor hunting. Subsequently, various religious or quasi-religious vegetarian doctrines may have arisen in an attempt to present an unwelcome necessity as a virtue.

No one knows exactly when our ancestors first acquired a strong taste for meat. Among our closest primate relatives, gorillas, orangutans and pygmy chimpanzees (bonobos) eat virtually no animal tissues of any kind but the common chimpanzee does include some meat in its diet. It is possible that the common ancestor of humans and chimpanzees ate significant amounts of meat or, alternatively, hunting behavior and meat-eating may have evolved independently in humans and chimpanzees. What is certain is that humans have been hunting and/or scavenging and eating meat for a very long time. An archaeological study of a tool-and-fossil-rich site with an age of 1.76 to 1.86 million years in Olduvai gorge, Tanzania, provides abundant evidence that early *Homo* had ready access to meat-rich carcasses of various African antelopes and other animals.[15] Carnivores such as lions or hyenas remove meat and marrow from bones by gnawing on them. Their teeth leave U-shaped grooves and round puncture holes. Early man cut meat from bones with stone tools that left V-shaped cuts and also smashed bones with hammer stones to extract marrow. Thus, the feeding activities of humans and conventional carnivores can be distinguished by careful microscopic study of the bones and bone fragments left behind. Further, U-shaped grooves superimposed on V-shaped grooves indicate that humans were often the primary predators in Tanzania nearly 2 million years ago, and that carnivores subsequently scavenged carcasses discarded by humans.

Similar work carried out in the Middle Awash Valley in Ethiopia indicates that early human ancestors were cutting flesh from bones and using hammerstones to break marrow bones from prey species that included antelopes of various kinds, pig-like animals, and ancient three-toed horses as early as 2.5 million years ago.[16] It may be that the early development of the use of tools as an aid to carnivory played a major role in the evolution of the reduced dentition and elaborate manipulatory abilities which are characteristic of humans.

Since most primates are largely herbivorous, it seems probable that the ancient common ancestor of humans and the living apes was also herbivorous. Since herbivores and carnivores display characteristic differences in the structure and function of the gastrointestinal tract, it would seem likely that the human gut would have gradually evolved from a herbivore-like pattern to a more carnivore-like pattern. Figure IV.1 illustrates some typical mammalian anatomical patterns.[17] Digestion in a typical carnivore such as the dog is largely dependent on enzymes that are released into the stomach and small intestine. Proteins are decomposed into amino acids, fats are decomposed into fatty acids. These products are then absorbed, mainly in the small intestine, distributed in the blood stream and used to provide energy or to synthesize new proteins and fats in various body tissues. Glucose, the major metabolic fuel of nervous tissue, is provided by the breakdown of glycogen, a polymer of glucose found in muscle and liver, and by synthesis (gluconeogenesis) from certain amino acids (alanine, aspartic acid, glutamic acid).

Digestion in herbivorous animals differs markedly from this pattern. Some of the major components of plants such as cellulose, hemicelluloses, and pectin cannot be hydrolyzed by any of the digestive enzymes that mammals can produce. However since some bacteria are able to produce enzymes that will decompose these giant organic molecules, herbivorous animals have evolved a symbiotic relation with them. Ruminant animals such as cattle or sheep have a greatly enlarged stomach consisting of four chambers: the reticulum, the rumen, the omasum, and the abomasum. The abomasum corresponds to the simple stomach of a carnivore: it begins the hydrolysis of protein by the enzyme pepsin in a strongly acidic environment. The other three chambers constitute a large fermentation tank in which a culture consisting of masticated plant material thoroughly mixed with saliva is kept at a high and constant temperature to favor the growth of many trillions of bacteria plus protozoa which feed on the bacteria. These micro-organisms synthesize proteins which can be digested in the abomasum and small intestine. Thus, ruminants actually live, not on plants themselves, but on a sort of yogurt-like material containing the bodies and metabolic products of vast numbers of bacteria and protozoa which they house in their stomach.

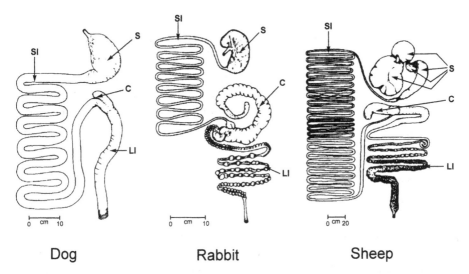

Dog   Rabbit   Sheep

Figure IV.1. Sketches of the gross morphology of the gastrointestinal system in the dog, rabbit and sheep (redrawn from Swenson, M.J. & Reece, W.O. (eds) *Dukes' physiology of domestic animals, 11th ed.*, Ithaca, N.Y. Cornell University Press, 1993). In all mammals the surface area of the small intestine is related closely to body size regardless of the customary diet.[18] Carnivora, such as dogs, usually have a small stomach, a small cecum, and small large intestine but in herbivores either the stomach is greatly enlarged, as in the sheep and other ruminants, or the large intestine and cecum are greatly enlarged, as in the rabbit. Since fermentation in the large intestine and cecum is a less efficient means of obtaining nutrition (absorption from the large intestine is rather poor) rabbits have adopted the habit of eating a part of their own feces in order to obtain additional nutrients on a second pass through the small intestine. The fecal pellets which are eaten differ from those that are excreted in the ordinary way in that they are soft, have a protective membrane, are specially formed in the cecum, and are eaten as they emerge from the anus. The eating of feces (coprophagia) occurs in many herbivorous mammals. For example, elephant calves eat large quantities of fresh adult elephant dung, a behavior that allows them to acquire the proper set of intestinal bacteria. These facts illustrate an important principle of evolution and natural history: Mother Nature has no shame. C, cecum; LI, large intestine; S, stomach; SI, small intestine.

In non-ruminant herbivores such as rabbits, horses, many rodents and some primates, the cecum and/or large intestine are greatly enlarged to provide a microbial fermentation chamber while the stomach has remained relatively small. Colobus monkeys and gorillas are examples of this type. Many primates eat a good deal of fruit and have a rather unspecialized type of gastrointestinal tract, but humans and cebus (capuchin) monkeys are unusual since they have a small stomach, a small cecum and a small large intestine. This is a pattern that approximates the one observed in most carnivores.[18] Since capuchin monkeys eat a great deal of animal matter (insects, small lizards, birds, etc) they, like humans, appear to have evolved a carnivore-like gastrointestinal system.

The evolution from a herbivore-type of gut to a more carnivore-type of gut in humans presumably required a rather long period of time. This agrees with archaeological evidence indicating that the ancestors of modern humans began eating significant amounts of meat a very long time ago. There is a third line of evidence that provides further support for this – the natural history of the tapeworm.

Tapeworms are parasitic flatworms which, when adult, live in the intestine of meat-eating mammals (the definitive host). An adult tapeworm has a head or scolex equipped with hooks and suckers to cling to the mucosa of the intestine. Behind the head is a neck which buds off a succession of reproductive units or proglottids. This results in a long segmented ribbon-like body structure which can grow to as much as several meters in length. Each proglottid contains both male and female sexual organs so that tapeworms can fertilize either themselves or a different individual. Ripe proglottids, filled with fertilized eggs, detach themselves from the tapeworm, are passed out with the feces, and disintegrate, releasing the eggs on the ground, grass, etc. If the eggs are then swallowed by a herbivorous animal of an appropriate species (the intermediate host) the eggs develop into larvae in the host's intestine. The larvae burrow into blood or lymph vessels and are carried to the host's muscles or other tissues where they form cysts or bladderworms which remain quiescent for long periods. If the herbivore host is then eaten by a carnivore, the larval tapeworms attach themselves to the wall of the intestine, grow into adults and repeat the cycle again.

Different species of tapeworm are adapted to specific predator-prey pairs and may have difficulty surviving in the tissues of other species. Thus *Taenia solium* and *Taenia asiatica*, the pork tapeworms, are adapted to the pig as the intermediate host and the human as the definitive host. Similarly, the beef tapeworm, *Taenia saginata*, is adapted to cattle as the intermediate host and to humans as the definitive host. If one could tell how long it took for these distinct species of human tapeworms to evolve, then one would know approximately when humans first began to feed regularly on the muscle tissue of pigs, cattle, or their relatives.

An attempt to do this by studying structural and biochemical differences in various species of tapeworms suggests that human tapeworms arose from tapeworms that infest large African cats (cheetahs, lions) hyenas, jackals or African hunting dogs. Early humans, from 90,000–1,710,000 years ago, or more, acquired these parasites by eating various prey animals which were also eaten by the carnivores. Long afterwards, when people domesticated cattle and pigs they transmitted the tapeworms to these animals.[19]

During the course of evolutionary history, metabolic processes tend to become adapted to an animal's habitual diet. The domestic cat, for example, has descended from a line of carnivorous felid or cat-like ancestors extending back at least 35 million years. It appears that during this time a number of metabolic pathways needed by herbivorous or partly herbivorous animals have disappeared in the cat.[20] Thus, most adult mammals, including humans, can survive indefinitely on a diet lacking the amino acid arginine because it can be synthesized in the body. Cats have lost this ability, presumably because a meat diet is rich in arginine, making such a synthetic pathway superfluous. An adult cat fed a single meal of an amino acid diet, complete except for arginine, rapidly developed high blood levels of ammonia associated with vomiting, muscle spasms, neurological symptoms and death. Similarly, cats are absolutely dependent on a dietary source of niacin or nicotinic acid which is abundantly present in meat, because they have largely lost the ability (present in humans) to synthesize this substance from the amino acid tryptophan. Cats have several other metabolic peculiarities. For example, most mammals can synthesize vitamin A from β-carotene, a substance occurring in plants. Since adequate levels of vitamin A are present in meat, cats, having no need of this particular pathway, no longer possess it. That is to say, in the remote past a mutant line of cat ancestors that lacked the dioxygenase enzyme responsible for converting β-carotene to retinal suffered no selective disadvantage and may even have had a slight advantage because they did not waste metabolic resources on an unnecessary bit of biochemistry.

It appears that domestic cats (and perhaps, other cat-like animals) are obligatory carnivores. It is really not possible for the lion to lie down with the lamb. Although humans are certainly less committed to an all-meat diet than cats, our species is not well adapted to a completely vegetarian diet. Plant proteins are less readily digested than animal proteins and are usually deficient in one or another amino acid. Thus, gliadin and gluten, the principal proteins in wheat, are deficient in lysine: zein, the principal protein in corn is deficient in tryptophan. Since we have little capacity to store amino acids, several of such deficient proteins must be eaten at each meal if no animal protein is eaten. We cannot avoid lysine and tryptophan deficiencies by eating wheat one day and corn the next day. Deficiency diseases such as kwashiorkor or pellagra are common in parts of the world where, for economic or religious reasons, people subsist on totally vegetarian diets. Such a diet also appears to be associated with an increased incidence of peptic ulcers.[21]

Vitamin B_{12} is a metabolic necessity which cannot be synthesized by mammalian tissues. Bacteria in the stomach and/or intestines of herbivorous animals synthesize B_{12} which is absorbed by the intestine and distributed to various tissues, providing a source of B_{12} for carnivores. Since humans lack sufficient intestinal bacterial fermentation to supply B_{12}, they can acquire this vitamin only by eating animal tissues (or by buying a supply from a modern drugstore). Consequently strict vegetarians and their breast-fed infants are likely to become deficient in B_{12}. In extreme cases, this may produce megaloblastic anemia, a condition characterized by abnormal red blood cell formation, plus gastrointestinal and neurological abnormalities.

Further, humans absorb the iron contained in hemoglobin at a higher rate (25–35%) than iron from other sources (2–20%). Additionally, meat contains a special "meat factor" that markedly increases iron absorption from non-hemoglobin sources such as vegetables, fruits and grains[21]. Consequently, a diet totally lacking in red meat may produce an iron deficiency anemia, especially in cases where the requirements of hemoglobin synthesis are high as in menstruating women (who lose about 35 ml of blood every month) or in pregnant women (who need extra iron to provide for a growing fetus). Since testosterone produces an increase an the formation of red blood cells, the onset of puberty in boys is also associated with an increased requirement for iron.

According to Washburn and Lancaster,[22] the fossil evidence indicates that human ancestors have been hunting and eating meat for at least 2.4 million years but that eating fish, shellfish and grinding seeds and nuts for food (many such products are indigestible to man without grinding, cooking, etc.) appeared only in the past few thousand years. Consequently, there may not have been sufficient time for the human gastrointestinal system to evolve effective adaptations to such foods. This may be related to the fact that many people develop a pathological condition of the mouth and intestine (sprue)[23] in reaction to ingestion of gluten (a protein found in wheat and other grains) and to the widespread occurrence of severe allergic reactions to fish, shellfish, peanuts, or tree nuts. Cow's milk, a food that would have been widely available only after the development of agriculture, is also a frequent cause of severe allergic reactions.[24] Similarly, many human adults are intolerant of lactose, the principal carbohydrate in milk. In a pre-agricultural world, only infants require intestinal lactase, an enzyme essential to the digestion of lactose.

In conclusion, if one examines the available evidence objectively, it appears that modern humans are well adapted to a largely carnivorous diet. An interesting perspective on this is provided by a study of the San or Bushmen, a hunting-gathering group of people living in the Kalahari Desert in South Africa.[25] In the Kalahari, rain falls in the December to April period and hunting is best from April to August. Meat accounted for a average caloric intake of 2,260 calories per person in July but during a considerable part of the year, the San were forced to subsist mainly on vegetable products. The body weight of adults over 20 years of ago reached an average maximum of 46 kg in June and July when meat was abundant but dropped to about 43.5 kg in January and February when mainly vegetable foods were eaten. Further, it appeared that 32 per cent of all births occurred in March and April, nine months after the time when meat intake and body weight were at their peaks. Good health and high fertility appear to be associated with meat-eating in the San people.

These observations may help illuminate an interesting aspect of human evolution. Carnivory provides access to a concentrated food that is easily digested and absorbed but it also carries a significant risk. Predators are notoriously susceptible to periods of famine if, for any reason, their habitual prey becomes scarce. Consequently many "carnivorous" animals such as foxes, coyotes, martens, etc., are quite willing to supplement their diets with fruit or corn.[26] Hunting and gathering aboriginal peoples all over the world seem to have evolved a life style in which men hunt and women gather vegetable foods.[27] Thus, in good times the family can enjoy steak while in bad times they can make do with roots and berries. Human ancestors evidently saw the wisdom of a maxim of modern-day investors: never put all your eggs in one basket.

It is controversial whether or not the diet of preagricultural man can serve as a guide to the selection of a healthful diet under modern conditions. There is much evidence that high blood levels of cholesterol and low density lipoproteins favour the development of atherosclerosis, leading to heart attack, stroke, and other disorders. Multiple factors including a hereditary predisposition, diabetes, obesity, lack of exercise, diet, the use of tobacco, etc., contribute to the development of atherosclerosis. Eskimos consuming the traditional fat meat diet have low levels of blood cholesterol and, although only limited data are available, no unusual levels of vascular disease. When Eskimos adopt the mixed diet characteristic of other segments of the North American population (including a large proportion of carbohydrates) they display increases in "obesity, cardiovascular disease, hypertension, and tooth decay" (see paper by H.H. Draper mentioned in Note #12).

A recent study in which obese American subjects were randomly assigned to either a high-fat, high-protein, low-carbohydrate diet or a low-calorie, high-carbohydrate, low-fat (conventional) diet concluded that over a six-month period the high-fat, high-protein, low-carbohydrate diet resulted in more weight loss than the conventional diet, a greater increase in blood levels of high-density lipoproteins, and a greater decrease in blood levels of triglycerides.[28] These effects are believed to decrease the risk of atherosclerosis. However, many subjects abandoned the carnivorous diet and many of the beneficial effects had disappeared within 12 months. Clearly, more extensive investigations are needed of the relative effects of these diets.

Modern humans display many anatomical and functional characteristics which can be understood as evolved adaptations to a hunting-gathering mode of life. Agriculture seems to have evolved only in the last 10,000 years or so and then only in certain parts of the world.[29] Many aboriginal peoples continued to live in a hunting-gathering mode of life well into the last two centuries. From a behavioral point of view such people differ very little from people living in technologically advanced societies. There is no doubt that living humans constitute a single though rather varied species.[30] Consequently, it is reasonable to assume that we are all children of the long paleolithic era of human life and that our behavior is fundamentally the behavior of a hunting-gathering mammal.

A striking characteristic which distinguishes humans from other predatory animals is that hunting is performed almost exclusively by males. In Murdock's survey[31] of 179 different human cultures, hunting was carried out exclusively by men in 166 cultures or tribes (92.7%) while women took part in hunting infrequently or in a subordinate capacity in 13 cultures. There were no cases in which women played the major role in hunting or even a role equal to the role of men.

It is interesting that in chimpanzees, as in humans, hunting is carried on primarily or exclusively by males but the meat obtained is shared with females and juveniles.[32] Since it is somewhat improbable that this unusual pattern of exclusively male predatory behavior would have evolved independently in two closely related species, it seems likely that the common ancestor of humans and chimpanzees was also an active hunter and that its hunting was carried on primarily by adult males. Similarly, the fact that chimpanzees hunt but do not scavenge already dead animals can be taken as evidence that scavenging never played an important role in human evolution.

There are numerous differences between modern men and women that can be understood as adaptations to a way of life in which men hunted over wide territories while women remained close to a temporary or permanent camp caring for the children, foraging nearby for vegetable foods and other materials, and working in the camp. Murdock reports that in a majority of human cultures such activities as gathering fruits, berries, nuts, herbs, roots, seeds, and fuel for fires are performed primarily by women. It is interesting that the gathering of shellfish, a prey which can neither flee nor defend itself, is also done primarily by women. An analogous situation occurs in chimpanzees in which the behavior of "fishing' for termites with a grass stem or twig occurs primarily in females.[33]

Although modern humans living in an urbanized society may have no experience whatsoever with hunting or gathering food and other materials in a natural environment they continue to display behavioral adaptations that probably evolved in remote prehistory.[34] Men, on average, are better than women in performing "spatial" tasks such as finding an efficient route from one place to another. Further, in finding their way about, women tend to rely heavily on visible landmarks while men rely more on abstract directions (north, east, south, west). For a hunter traveling long distances over unfamiliar or only partly familiar terrain, landmarks are of limited utility but a knowledge of direction is essential.

The adaptation of human males to traveling long distances while hunting may also be revealed in the differing tolerance of modern men and women to long-continued vigorous exercise such as running. A substantial proportion of female athletes develop a triad of disorders which includes reproductive disturbances (amenorrhea, failure to ovulate), osteoporosis, and eating disorders (anorexia, bulimia).[35] Comparable disorders rarely occur in male athletes.

Men are, on average, much more accurate than women in throwing darts or balls at a target, an ability that probably evolved in relation to our species long use of spears and other thrown missiles in hunting.[36]

Women, on the other hand, are superior to men on tasks involving fine motor control and accurate discrimination of small nearby objects,[34] abilities that would be useful in such ancient feminine pursuits as gathering seeds or berries, making baskets, etc. Women also seem to be better than men in detecting changes in facial expression, tone and posture that may indicate hostility, sexual attraction, etc. Such abilities would be of great value in detecting and resolving conflicts in a close-knit social group.

More generally, the transition from the vegetarian diet characteristic of most primates to a partially or largely carnivorous diet is likely to have led to the evolution of extensive changes in the social behavior of ancestral hominids.[37] Among mammals, males are often not well integrated into whatever social grouping a species may possess. In solitary mammals, such as most cats and many rodents, the male and female associate only during a brief mating period and the female raises the young alone. In some other species, such as the North American wapiti or elk, males avoid females and young during much of the year but during the annual rut males attempt to gather together and control a harem of females while aggressively driving off rival males. In such primates as baboons and macaques a similar pattern of male aggression and dominance prevails except that males and females remain together throughout the year.

Since high levels of intraspecific aggression will tend to interfere with co-operative group action, social carnivores such as wolves and African hunting dogs, which hunt in packs, have evolved various means of living together peacefully.[38] Pack hunting has several advantages, allowing a group of predators to attack successfully large prey which could not be captured by a single predator, and allowing one or more predators to drive prey toward conspecifics who lie in wait. The selective advantages of pack hunting would tend to promote effective communication among pack members and the evolution of mechanisms for reducing intra-group conflict. Thus, social carnivores such as wolves and African hunting dogs share their food without quarreling and have reduced conflicts over sexual opportunities. In a wolf pack, mating is usually confined to a single male and a single female. Other adults in the pack remain celibate and there is little or no sexually-inspired conflict. In gorillas, a single dominant male controls a harem of females but chimpanzees have a promiscuous mating system in which several males mate with an estrus female. There are prominent signs of estrus in female chimpanzees (conspicuous red swollen vaginal labia and circumanal tissue)[39] and mating does not occur during non-estrus periods. In humans, visible signs of estrus have disappeared, mating can occur at any time during the menstrual cycle, and although our species could fairly be described as somewhat polygynous,[40] there is a far greater tendency to form long-term monogamous heterosexual pairs then is the case in our nearest relatives, chimpanzees and gorillas. Concealment of estrus and an increased tendency toward monogamy (perhaps facilitated by the liberation of copulation from a purely reproductive function) may have evolved as a means to greater social harmony and co-operation. Presumably this should be regarded as an example of convergent evolution in which ancestral hominids developed social behaviors which are comparable in many respects to those prevailing among social carnivores such as wolves.

Food-sharing behavior provides another example of human evolution converging with the pattern found in social carnivores. An adult male wolf will bring meat to its den to supply its pups and the lactating mother. A similar pattern occurs in the red fox. Among non-human primates the sharing of food is uncommon or non-existent except for the tendency of male chimpanzees to share meat with other group members. Although this behavior may possibly have a common evolutionary origin with food-sharing in humans it is apparent that it is enormously more developed in humans than in chimpanzees. In hunting-gathering societies both men and women collect food far in excess of immediate personal need, transport it to a home base and share it communally with others. Food-sharing results not only in a more efficient use of resources but also contributes to the survival of sick or injured individuals who are temporarily unable to find food for themselves.

In conclusion it appears that during the course of human evolution a quadrupedal ape-like ancestor became bipedal, developed a large brain, acquired extensive tool-making and tool-using skills, acquired an unparalleled ability to communicate using gestures and audible speech, became increasingly dependent on meat as a food, developed a common but by no means universal tendency to form long-term male-female pairings, and developed an increased capacity for complex social organization. Some features of human evolution such as a reduction in the size of the jaws, the canines and other teeth may have been secondary consequences of the use of tools in fighting or predatory attack and in the preparation of food by grinding, slicing and cooking. Manipulatory and constructive abilities would have resulted in a selective advantage, not only for their use in tool making, hunting and the preparation of food, but also because they are essential in making shelters and clothing. Without such abilities hairless humans could not have colonized the colder regions of the world.

The extraordinary exploratory and manipulatory abilities of humans which led, eventually, to the development of science and technology may also owe their existence to behaviors that evolved in a primate that became a carnivore. S.E. Glickman and R.W. Sroges[41] placed various novel objects (lengths of steel chain, pieces of wood dowel, rubber tubing and a crumpled piece of white paper, each presented in succession) in the cages of various zoo animals. In general, they observed that carnivorous mammals (various canids, felids, mustelids, and procyonids) reacted to these objects by vigorous examination, touching and grasping them and knocking them about while herbivorous animals such as various rodents and marsupials reacted minimally or ignored the objects. Among primates, the entirely herbivorous *Colobinae* displayed little reactivity while the more generally omnivorous *Cercopithecinae* (including baboons and rhesus monkeys) reacted very vigorously. Thus, what is often termed the human quest for knowledge may have evolved because exploratory and manipulatory abilities were advantageous in an animal adopting a tool-assisted carnivorous life style.

Similarly, it has been pointed out by many observers of wildlife on the African plains that the different species of herbivore take little interest in one another. Wildebeest, zebras, and gazelles graze side by side with little or no interaction. In contrast, carnivorous animals take an acute interest in other species. It may appear paradoxical, but it is very likely that modern humans' interest in such pursuits as bird-watching or wildlife photography has its origin in adaptations to a life of hunting and carnivory.

Although evolutionary hypotheses of this type are easy and pleasant to imagine, it has been very difficult to identify the selective factors that may have led to the evolution of different distinctive human characteristics.[42] However, for many purposes it is sufficient to know what evolved even though we may not know exactly why it evolved.

Survival in the modern world. If it is really true that humans are best adapted to a Paleolithic lifestyle of hunting, gathering, and living in small groups, one can but wonder how we mange to survive in modern urban environments which are quite unnatural in both physical and social aspects. Experience with keeping wild animals in zoos has taught us that many species have special requirements with respect to water, type of earth, type of vegetation, light levels, etc. Such habitat preferences appear to be at least partly instinctive. It may be that the persisting appeal, in modern humans, of outdoor activities such as walking in a park, going on picnics, camping out, etc., is an indication of an innate attraction to such things as earth, rocks, water, green vegetation, and opportunities to view wild life (or animals in zoos). The consequences for human behavior of deprivation of these things can only be guessed at. One obvious consequence is that modern urban people typically have a very poor knowledge of the natural world, knowing little or nothing about the properties of different rocks or plants, animal behavior, the seasons, phases of the moon, etc. Whether isolation from the natural world contributes in any way to behavioral pathology in an urban environment is quite unknown. It is fairly obvious, however that life in a modern urban environment contributes to obesity and poor physical condition because it provides very little opportunity for strenuous physical exercise.

The modern urban world restricts or prevents many activities which are natural to humans. In a hunter-gatherer society, children of different ages play and work together, a situation that results in much learning by the younger children from the examples set by the older ones. It appears that many rhymes, songs, games, as well as more formal skills and knowledge about life are transmitted from child to child in this way without adult intervention. Large modern urban schools, designed on an industrial model, interfere with this process by rigid separation of children into grades based primarily on age. Similarly, the tendency to hive off elderly people into "retirement communities" deprives children of the opportunity to see first hand the full cycle of human life, the inevitability of death and the possibility of a point of view differing from one's own. Further, in a traditional society, children can observe directly the various tasks undertaken by their parents, allowing them excellent opportunities to acquire the skills and knowledge essential for successful adult life. In modern urban homes in which the father (and probably the mother as well) is absent most of the day, there is no opportunity for children to observe and learn from the work of their parents. It may be that television and computers are less than adequate as substitute means of teaching because they fail to provide a living role model with which a child can interact. These factors may increase the difficulty of making a successful transition from childhood to adult life.

From puberty onwards, young men in traditional societies compete and "show off" in games, hunting, or fighting to establish themselves in as high a social rank as possible. This behavior promotes alliances useful in later life and may provide mating opportunities. Women everywhere are attracted to successful high status men. (Consider the "groupies" who attach themselves to rock stars or sports stars.) The modern urban world provides no opportunities for demonstrating prowess in hunting, and discourages violent competition between young men, except for those few who have special aptitudes for professional sports. Paradoxically, this may increase urban violence because many young men are attracted to youth gangs which engage in high-risk criminal activity, daring one another to more and more excessive behavior in competitive attempts to gain social status.

Among people living in small hunter – gatherer groups, the perpetrator of an anti-social act is readily identified and may be subjected to sanctions or punishment of some sort. In the vast anonymous crowd in a modern urban centre, the perpetrators of antisocial acts may entirely escape detection, a fact which may play a role in the appearance of serial killers, serial rapists, pedophiles who lure young children with candy, etc.

Finally, although it seems clear that the development of civilization in the past ten thousand years or so has been of enormous benefit, we must also consider its costs. The scientific and technical knowledge that permits mechanized agriculture, modern medicine and such amenities as paved roads, electrical power, jet aircraft, and espresso coffee, also permits aerial bombing, the machine gun, and chemical and biological warfare. The social organization and communications that permit modern countries to enjoy unprecedented levels of peace and prosperity also permit concentration camps, mass torture and organized genocide. We are leaving our descendents much scope for improvement.

Notes

1. Darwin, C. (1998). *The descent of man*, Amherst, New York: Prometheus Books (first published, 1871).
2. A good comprehensive discussion of ancient hominid fossils and artefacts can be found in: Wolpoff, M.H. (1999). *Paleoanthropology*, 2[nd]ed., Boston, MA: McGraw-Hill. An interesting synopsis of the history of civilization has been provided by: Wright, R. (2004). *A short history of progress*, Toronto, Ontario: House of Anansi Press, Inc.
3. Data on brain volumes in apes, modern humans and human ancestors can be found in: Semendeferi, K., and Damasio, H. (2000). The brain and its main anatomical subdivisions in living hominids using magnetic resonance imaging. *Journal of Human Evolution*, 38: 317–332, and in Tobias, P.V. (1988). The brain of *Homo habilis*: A new level of organization in cerebral evolution. *Journal of Human Evolution*, 16: 741–761.
4. Leakey, M.D., and Hay, R.L. (1979). Pliocene footprints in the Laetoli beds, northern Tanzania, *Nature, London*, 278: 317–323.
5. White, T.D., and Suwa, G. (1987). Hominid footprints at Laetoli: facts and interpretation. *American Journal of Physical Anthropology*, 72: 485–514.
6. The classic account of the behavior of wild chimpanzees is: Goodall, J. (1986). *The chimpanzees of Gombe: patterns of behavior.* Cambridge, MA: Harvard University Press.
7. Napier, J.R. and Napier, P.H. (1985). *The natural history of primates.* London: British Museum (Natural History), p. 53.
8. Schmid, H. (1971). Foot skills in children with severe upper limb deficiencies. *The American Journal of Occupational Therapy*, 25: 159–163. Smithells, R.W. (1973). Defects and disabilities of thalidomide children. *British Medical Journal*, 1: 269–272. The comment quoted is from Smithells' paper.
9. Lack, D. (1983). *Darwin's finches*. Cambridge, U.K.: Cambridge University Press, pp 58–59.
10. Smith, S.M. (1972). The ontogeny of impaling behaviour in the loggerhead shrike, *Lanius ludovicianus L. Behaviour*, 42: 232–247.
11. Abrams, H.L., Jr. (1987). The preference for animal protein and fat: A cross-cultural survey. In: M. Harris and E.B. Ross (eds.) *Food and evolution*, Philadelphia: Temple University Press, pp. 207–223.

12. Draper, H.H. (1977). The aboriginal Eskimo diet in modern perspective. *American Anthropologist, 79*: 309–316.
    Lieb, C.W. (1926). The effects of an exclusive long-continued meat diet, based on the history, experiences and clinical survey of Vilhjalmur Stefansson, Artic explorer. *Journal of the American Medical Association, 87*: 25–26.
    Lieb, C.W. (1929). The effects on human beings of a twelve month's exclusive meat diet. *Journal of the American Medical Association, 93*: 20–22. Stefansson, V. (1960). Food and food habits in Alaska and northern Canada. In: I. Galdston (ed.) *Human nutrition: Historic and scientific, Monograph III. The New York Academy of Medicine*, New York: International Universities Press, pp. 23–60.

13. Innis, H.A. (1962). *The fur trade in Canada.* Toronto: University of Toronto Press, p 235 ff. (first published 1930). The manufacture of pemmican is described by: Steele, S.B. (1973). *Forty years in Canada: Reminiscenses of the great north-west*, Toronto: Coles Publishing Company, see pp. 95–96 (first published in 1915).

14. Martin, P.S., and Klein, R.G. (eds, 1984). *Quaternary extinctions: a prehistoric revolution.* Tucson, Arizona: The University of Arizona Press.
    Although there is a good deal of circumstantial evidence that human hunting was a primary cause of the extinction of many animals, it has not been possible to prove this in a rigorous way and show that other factors such as disease or changes in climate were not responsible.

15. Oliver, J.S. (1994). Estimates of hominid and carnivore involvement in the FLK *Zinjanthropus* fossil assemblage: some socioecological implications. *Journal of Human Evolution, 27*: 267–294.

16. de Heinzelin, J., Clark, J.D., White, T., Hart, W., Renne, P., Wolde Gabriel, G., Beyene, Y., and Vrba, E. (1999). Environment and behavior of 2.5-million-year old Bouri hominids. *Science, 284*: 625–629.

17. A good source-book for information about digestion and other aspects of physiology in various domestic animals is: Swenson, M.J., and Reece, W.O. (eds). *Dukes' physiology of domestic animals* 11th ed. Ithaca, N.Y.: Cornell University Press, 1993.

18. Martin, R.D., Chivers, D.J., MacLarnon, A.M., and Hladik, C.M. (1985). Gastrointestinal allometry in primates and other mammals. In: W.L. Jungers (ed.) S*ize and scaling in primate biology*, New York: Plenum Press, pp. 61–89.

19. Hoberg, E.P., Alkive, N.L., de Queiroz, A., and Jones, A. (2001). Out of Africa: origins of the Taenia tapeworms in humans. *Proceedings of the Royal Society of London B, 268*: 781–787.

20. Morris, J.G., and Rogers, Q.R. (1982). Metabolic basis for some of the nutritional peculiarities of the cat. *Journal of Small Animal Practice, 23*: 599–613.

21. A good discussion of dietary deficiency diseases can be found in: E. Braunwald, S.L. Hauser, A.S. Fauci, D.L. Longo, D.L. Kasper, and J.L. Jameson (eds.) *Harrison's principles of internal medicine*, 15th ed. 2001. New York: McGraw-Hill, pp. 451–469. A more biochemically oriented discussion is presented in: A. White, P. Handler, E.L. Smith, R.L. Hill and I.R. Lehman (1978). *Principles of biochemistry*, New York: McGraw-Hill, pp. 1320–1379. Dietary absorption of iron is discussed by Monsen, E.R., Hallberg, L., Layrisse, M., Hegsted, D.M., Cook, J.D., Mertz, W., and Finch, C.A. (1978). Estimation of available dietary iron. *American Journal of Clinical Nutrition, 31*: 134–141.

22. Washburn, S.L., and Lancaster, C.S. (1968). The evolution of hunting. In: R.B. Lee and I. DeVore (eds) *Man the hunter.* Chicago: Aldine-Atherton, pp. 293–303.

23. Sprue is discussed in the references in note #21.

24. Ewan, P.W. (1997). Anaphylaxis. *British Medical Journal, 316*: 1442–1445.

25. Wilmsen, E.N. (1978). Seasonal effects of dietary intake on Kalahari San. *Federation Proceedings, 37*: 65–72.

26. A classic discussion of the larger implications of predation can be found in: Errington, P.L. (1967). *Of predation and life.* Ames, Iowa: Iowa State University Press.

27. Coon, C.S. (1971). *The hunting peoples.* Boston: Little, Brown and Company. Also see: Bicchieri, M.G . (ed., 1972). *Hunters and gatherers today.* New York: Holt Rinehart and Winston, Inc.

28. Foster, G.D., Wyatt, H.R., Hill, J.O., McGuckin, B.G., Brill, C., Mohammed, B.S., Szapary, P.O., Rader, D.J., Edman, J.S., and Klein, S. (2003). A randomized trial of a low-carbohydrate diet for obesity. *The New England Journal of Medicine, 348*: 2082–2090.

29. Struever, S. (ed.) (1971). *Prehistoric agriculture,* Garden City, New York: The American Museum of Natural History, The Natural History Press.

30. Eibl-Eibesfeldt, I. (1989). *Human ethology,* New York; Aldine de Gruyter.

31. Murdock, G.P. (1965). *Culture and society,* Pittsburgh: University of Pittsburgh Press, pp. 308–310.

32. Teleki, G. (1973). The *predatory behaviour of wild chimpanzees,* Lewisburg: Bucknell University Press.

33. McGrew, W.C. (1992). *Chimpanzee material culture,* Cambridge, U.K.: Cambridge University Press.

34. Kimura, D. (2000). *Sex and cognition,* Cambridge, Massachusetts: MIT Press.

35. Ireland, M.L. and Nattiv, A. (eds. 2002). *The female athlete,* Philadelphia, Pennsylvania: Elsevier Science.

36. An excellently preserved spear, 2.3 m in length and made from an individual spruce tree that had been debarked and shaped like a modern javelin with a heavy front end, a sharp tip, and a long tapering tail was found in German geological deposits estimated to be about 400,000 years old [Thieme, H. (1997). Lower paleolithic hunting spears from Germany. *Nature, 385*: 807–810]. Such a spear, having good aerodynamic properties which would cause it to travel long distances in a point-first orientation when thrown, suggest that spear-making was already an ancient and well-understood art nearly half a million years ago. Associated with this spear were other wooden tools, worked flints, and over 1,000 large mammal bones, many bearing the marks of butchering.

37. Etkin, W. (1954). Social behaviour and the evolution of man's mental faculties. *The American Naturalist, 88*: 129–142.

38. Schaller, G.B., and Lowther, G.R. (1969). The relevance of carnivore behaviour to the study of early hominids. *Southwestern Journal of Anthropology, 25*: 307–341.

39. Short, R.V. (1979). Sexual selection and its component parts, somatic and genital selection, as illustrated by men and the great apes. *Advances in the study of behaviour, 9*: 131–158.

40. Daly, M., and Wilson, M. (1978). *Sex, evolution and behaviour,* Belmont, California: Wadsworth Publishing Co.

41. Glickman, S.E., and Sroges, R.W. (1966). Curiosity in zoo animals. *Behaviour, 26*: 151–188.

42. A non-exhaustive list of hypotheses to account for bipedalism, loss of body hair, and various other characteristics of modern humans can be found in the following papers:

    a) Cant, J.G.H. (1981). Hypothesis for the evolution of human breasts and buttocks. *American Naturalist, 117*: 199–204.

    b) Ebling, J. (1985). The mythological evolution of nudity. *Journal of Human Evolution, 14*: 33–41.

    c) Hunt, K.D. (1994). The evolution of human bipedality: Ecology and functional morphology. *Journal of Human Evolution, 26*: 183–202.

    d) Jablonski, N.G., and Chaplin, G. (1993). Origin of habitual terrestrial bipedalism in the ancestor of the *Hominidae. Journal of Human Evolution, 24*: 259–280.

    e) Kushlan, J.A. (1985). The vestiary hypothesis of human hair reduction. *Journal of Human Evolution, 14*: 29–32.

f) Rodman, P.S., and McHenry, H.M. (1980). Bioenergetics and the origin of hominid bipedalism. *American Journal of Physical Anthropology, 52*: 103–106.

g) Sheets-Johnstone, M. (1989). Hominid bipedality and sexual selection theory. *Evolutionary Theory, 9*: 57–70.

h) Sanford, C.B. (1999). *The hunting apes: meat eating and the origins of human behaviour.* Princeton, New Jersey: Princeton University Press.

i) Wheeler, P.E. (1984). The evolution of bipedality and loss of functional body hair in hominids. *Journal of Human Evolution, 13*: 91–98.

j) Wheeler, P.E. (1985). The loss of functional body hair in man: the influence of thermal environment, body form, and bipedality. *Journal of Human Evolution, 14*: 23–28.

k) Wheeler, P.E. (1993). The influence of stature and body form on hominid energy and water budgets: A comparison of *Australopithecus* and early *Homo* physiques. *Journal of Human Evolution, 24*: 13–28.

V. Human instinctive behavior

Anyone who takes even a casual interest in the behavior of wild or domestic animals cannot fail to notice that different species have their own characteristic ways of doing things. Gilbert White,[1] an eighteenth century English naturalist, noted that although field mice, squirrels, and nut-hatches all feed on hazel nuts, each species has its own special way of opening them. White believed that many such behaviors are largely instinctive, that is, they are inborn, innate, or unlearned. He pointed out, for example, that a viper's characteristic movements and the tendency to bite are present in fully developed form in young which have no fangs as yet and which have been surgically removed from their mother's abdomen before birth (vipers are ovoviviparous, i.e. the eggs hatch in the mother's oviducts and the young are born alive). It may be added that young vipers, delivered into the world in this way, would have had no opportunity to learn their characteristic defensive behavior.

It is likely, as was already noted by Gilbert White, that human behavior is also at least partly instinctive. Much of human behavior is quite independent of particular cultures or individual experience including: (a) bipedal locomotion; (b) extensive use of the forelimbs to manipulate objects; (c) some form of speech; (d) a varied and finely differentiated repertoire of facial movements; (e) a complex system of social behavior based on individual recognition and involving (usually) rather marked differences in behavior between the two sexes; and (f) patterns of reproductive and parental behavior which are unusual, possibly unique, among primates. Common observation suggests that much of this is species-specific and instinctive. In many households, kittens, puppies and children all grow up together, eating the same food in many cases, all played with and all spoken to by the adult humans who live there. Nonetheless, the kittens grow up to behave like cats, the puppies grow up to be dogs and the children are transformed into adult humans. Even children who are horribly abused usually develop articulate speech and the rudiments of normal human behavior.

The study of instinctive behavior was transformed from casual observations and anecdotes to a systematic scientific field primarily by the efforts of Konrad Lorenz (1903–1989) an Austrian zoologist, and Nikolaas Tinbergen (1907–

1988) a Dutch zoologist who moved to a position at Oxford University in England in 1949.

Among the concepts developed by Lorenz and Tinbergen was the idea that instinctive behavior, in general, involves the activation of "fixed action patterns" (characteristic taxon-specific items of behavior) by specific environmental stimuli (termed "releasers" or "releasing stimuli").[2] For example, a male stickleback fish (*Gasterosteus aculeatus*) will, in its spring-time mating period, attack another stickleback which has a red underside (present only in males). The red colored patch is essential since a crude wooden cigar-shaped model will suffice to elicit an attack provided that its underside is painted red but an accurate fish model lacking the red patch is ignored. Lorenz and Tinbergen further assumed that once a fixed action pattern (such as the attack behavior of a stickleback) had been triggered it would carry on to the completion without feedback from the movements already performed. Another characteristic of many instinctive behaviors, which was emphasized by Tinbergen, is the tendency to react in an automatic and unintelligent manner. For example, the oyster catcher (*Haematopus ostralegus*) will attempt to incubate a giant artificial egg in preference to its own eggs when given a choice. The giant egg is evidently a more effective eliciting stimulus for incubation than the bird's own eggs.

One of the intriguing concepts proposed by Lorenz and Tinbergen was the idea that certain anatomical structures have evolved, together with an associated behavior, in order to send signals to other animals, especially those of the same species. The signals, then, elicit some instinctive act in the recipient.[3] Thus, many species of birds have elongated tail feathers, ruffs, crests, bright colors, etc., together with specialized movement patterns that show off the anatomical features in a most impressive manner. The elaborate tail fanning and strutting courtship displays of male grouse, peacocks and turkeys provide examples. It is presumed that these anatomical and behavioral features have evolved because they are effective elicitors of mating behavior in females.

In humans too, some morphological features and the behaviors that display them to advantage may have evolved because they elicit instinctive reactions in other humans.[4] The beard may serve as a sexual signal in human males, (although the actual function of the beard, if any, remains to be demonstrated) and the distinctive mobile eyebrows may serve as social signals in both sexes. Pubertal enlargement of the hips and breasts in girls is a uniquely human adaptation, an indication of sexual maturity. In other mammals the mammary glands become enlarged only during lactation, not throughout the year as is the case in humans, and the hips are generally not enlarged at all. It may be that, in parallel with the evolution of these morphological features in women, together with relevant display behaviors (side-to-side movement of the pelvis during walking, the wearing of bustles, tight jeans, padded brassieres, etc.) the male human brain developed neural circuitry capable of reacting to such signals by the activation of courtship behavior.

Lorenz's and Tinbergen's proposal that behavior can be inherited in much the same way that anatomical, biochemical, or physiological features can be inherited aroused strong opposition from people who believed that behavior, especially in humans, is largely or totally a matter of learning and culture. Although this debate still continues in some quarters (radical feminists, for example, insist that any differences that there may be in the behavior of men and women are due entirely to cultural influences), it has become clear that instinctive behavioral tendencies interact with the effects of experience in a complicated manner.[5] This interaction is well illustrated by a classic study on the development of predatory behavior in cats.

Most people who keep domestic cats as pets will have noted that a mother cat will carry home mice, chipmunks, sparrows, or other small prey, crippled but still alive, which are offered to the kittens to play with. The mother lies nearby, watching attentively as the kittens chase, catch, strike and bite the prey. If the prey animal attempts to escape or fights back, attempting to bite a kitten for example, the mother attacks it immediately, biting it severely enough to cripple it further but not enough to kill it. The mother will also bring home prey which she eats as the kittens watch. Similar behaviors occur in larger cats such as tigers and cheetahs.[6]

It would appear that mother cats engage in a form of active teaching of their kittens. Is the predatory behavior of cats then truly instinctive or is it dependent on learning and the transmission of a hunting culture?

Zing Yang Kuo, in a classic study published in 1930,[7] raised groups of kittens under varying conditions prior to exposing them to a rat or a mouse in repeated 30 min tests. Nine kittens in a group of 20 raised in isolation (separated from the mother at weaning, no experience with rats or mice whatever) made one or more kills before the age of 4 months. In contrast, 18 kittens in a group of 21 who had witnessed their mother killing a rat or a mouse in an adjacent cage, made kills before the age of 4 months. In a group of 18 kittens raised alone in a cage with a rat or a mouse companion from the age of 6–8 days (the mother was removed from the cage every day when the rat or mouse was present) only 3 ever made a kill in Kuo's tests.

Kuo's experiments indicate that killing mice and rats is partially instinctive in cats. Nearly half (9/20) of a group of kittens that had never witnessed a predatory attack and had no experience with rats or mice made one or more kills. Three kittens in a group of 18 that had been habituated to rats and mice from an early age similarly killed at least one rat or mouse.

It is relevant to note that if Kuo had used non-carnivorous animals such as guinea pigs or goats as experimental subjects he would have observed no rat or mouse killing whatsoever. Cats, unlike guinea pigs or goats, have an inherited disposition to attack, kill, and eat small mammals. This disposition, however, becomes fully developed only as a result of experience. Most of Kuo's kittens (18/21) that had watched their mother killing a rat or mouse in an adjacent cage became successful predators themselves. However, some kittens (3/21) did not become successful predators even after watching a demonstration of predation from a distance. It is possible that these three kittens would also have become successful predators if they had received the complete natural training course, which includes playing with crippled prey, but Kuo did not test this.

There are data suggesting that the role of experience in instinctive behavior is, at least in part, a matter of determining which stimuli will become effective elicitors of the behavior and that the instinctive behavioral performance itself develops in its natural form even in the absence of relevant experience. This is shown, for example, by studies of the effect of stimulation of the lateral hypothalamus in cats.[8] Electrical stimulation of a limited region of the lateral hypothalamus elicits what has been termed a "quiet biting attack" in which the cat approaches a stimulus rat with the body lowered and the neck extended. The forepaws are used to seize and hold the rat which is then quickly killed by biting. Such stimulus objects as a plastic sponge or a furry toy dog are less effective in eliciting attack during hypothalamic stimulation than a live rat. Presumably the visual and other stimuli afforded by the prey objects summate with the hypothalamic stimulus to activate the brain circuitry involved in predatory attack. Live rats are especially effective stimuli. Cats that had been reared in social isolation from the age of 5 days reacted to lateral hypothalamic stimulation in much the same way as normally reared cats in such tests except for a tendency to attack the inanimate stimulus objects more readily than normally reared cats would do. Since we know from Kuo's results that most cats reared in isolation will not spontaneously attack rats and mice, we can conclude that the predatory behavior pattern of cats develops normally in the absence of relevant experience but that isolation rearing makes this behavior less readily elicitable by natural stimuli.

Many human behaviors which have a characteristic taxon-specific form or pattern, nonetheless display great variability with respect to the effective eliciting stimuli. Children who are born blind and deaf have a much reduced opportunity of mimicking the behavior of other humans but they, nevertheless, display essentially normal patterns of smiling, laughing, pouting, crying and temper tantrums.[9] On the other hand, as one might expect, blind children cannot voluntarily pose happy or sad faces as well as normal children.[10] In normal humans, as everyone knows, there is considerable variation around the world in the stimuli that will elicit "genuine" (instinctive) smiling, laughing, pouting, crying and temper tantrums or the voluntary (posed) forms of such behaviors.

The conclusions suggested by these and other studies of instinctive behavior in mammals are that: (1) many distinctive behavior patterns develop sponta- neously without much influence from special training or culture; (2) experience has a major effect in determining which stimuli will be effective in eliciting a given behavior pattern; and (3) complex behavior patterns may be refined and further developed in various ways as a result of experience.

Neuroanatomical observations on the effect of environmental experience on the brain are broadly consistent with this pattern. The brains of rats, cats, guinea pigs, goats, etc., differ, not only in size and shape, but also in the patterns of internal connectivity of the neurons.[11] These innate differences in development and neural connectivity are, no doubt, responsible for the differing behaviors characteristic of these species. Animals that have been raised under different conditions or exposed to different learning situations show differences in the microstructure of the brain (the dendritic bush of specific neurons may be enlarged or reduced; patterns of synaptic connectivity may be altered) but the large scale features of the brain are little affected. Rats reared in a complex environment permitting many different experiences, have a cerebral cortex that is about 4% heavier than rats reared in a small cage in an unchanging environment. Thus, environmental stimulation and opportunity to learn has an effect on brain development but the effect is rather small.[12] No amount of training will make a cat brain look like a rat or rabbit brain just as no amount of training will make a cat behave like a rat or a rabbit.

Human speech provides a clear example of the cooperative influence of hereditary and experiential factors in the development of behavior. Speech is clearly a species-specific or instinctive characteristic of humans.[13] No other animal spontaneously arranges a limited repertoire of sounds and/or gestures into an almost infinite number of sequential patterns which convey differing propositions. Nonetheless, it is also clear that speech is a learned behavior. Children acquire the speech of the adults among whom they grow up and the mutually incomprehensible languages spoken around the world number in the thousands.

There is clear evidence that a distinct neural apparatus for the control of vocalization and gesture has evolved in the human brain. Long experience with the effects of localized ischemic brain damage (stroke) in humans has shown that the ability to speak can be disturbed or abolished by injury to several different regions of the neocortex, usually on the left side of the brain.[14] Comparable lesions in the monkey have no effect whatever on the various vocalizations of these animals.[15] The conclusions suggested by the effects of brain lesions are confirmed by the effects of localized electrical stimulation of the intact brain. In conscious humans, stimulation of the wide regions of the neocortex elicits either or both: (1) a sustained vocalization (usually a prolonged vowel sound); or (2) interference or blockade of normal speech.[16]

In contrast, electrical stimulation of the neocortex in monkeys or cats generally produces no vocalization at all although vocalization is readily elicited by stimulation of a variety of subcortical structures such as the amygdala, hypothalamus and central grey.[17]

Therefore, it appears to be the case that the evolution of language in humans was associated with the appearance of a neocortical control of vocalization which is not found in non-human mammals. Presumably this cortical control apparatus, established initially by developmental processes under the control of the genome, can be developed and modified by experience thereby making possible the great diversity of human languages which occur around the world.

Human speech appears to be a specialized instinctive behavior which is relatively independent of brain size or overall intellectual development. This may be illustrated by the syndrome of microcephaly, a congenital failure of the brain to develop normally.[18] Adult microcephalic patients may have a brain weight of 300 grams or even less (normal human brain weight is about 1400 grams), which is roughly equivalent to the brain weight in chimpanzees. Apart from small size, the microcephalic brain often has a normal or nearly normal appearance. If the brain weight is below about 500 grams, the patients are invariable idiots but, nonetheless, they display many normal human behaviors. They walk bipedally, use the hands to manipulate objects, laugh, cry, have temper tantrums, display many normal facial expressions, may dance or play music, and can speak in simple sentences, although they are usually incapable of carrying on a conversation. The fact that human microcephalic patients walk bipedally and speak in a limited fashion but chimpanzees and gorillas do not indicates that speech and bipedal locomotion are dependent on particular patterns of neural connections in the brain rather than on brain size.

Human speech, unlike many of the vocalizations of other mammals, is an operant or voluntary behavior controlled by rewards and punishments.[19] Attempts to train non-human primates to utter various vocalizations to obtain food, for example, have generally failed[20] but young children readily acquire the sequences of sounds that induce their parents, or other adults, to offer food, pick them up, etc. In the past half century or so, there has been much acrimonious and pointless debate about whether the rules of language are learned or innately specified. The truth is that all behavior, including speech, depends on neural circuits that develop under the guidance of the genome. This basic circuitry can then be refined and further developed to some extent by experience. In the case of speech, an important aspect of this experience is that words have social consequences. People are usually helpful if spoken to politely in a language they are familiar with. This encourages speakers to conform to the conventional norms of the local dialect.

One may assume that, since exposure to Russian leads to an ability to speak Russian, while exposure to English leads to an ability to speak English, that these differing experiences result in the establishment of different patterns of connectivity among cortical neurons. If our knowledge of cortical microstructure were sufficiently detailed, it should be possible, in principle at least, to determine which languages were spoken during life by study of cortical tissue removed soon after death.

It is clear that not only specific sensori-motor reactions, but also the overall organization of behavior in different species is controlled mainly by hereditary factors. Consider the organization of social and reproductive behavior in different mammals. In North American elk or wapiti (_Cervus canadensis_), for example, adult males (bulls) live alone or in company with other bulls, avoiding all contact with adult females (cows) and young (calves) during most of the year. In summer and autumn, in response to a light-controlled hypertrophy of the testis and the release of high levels of testosterone, the bulls develop bony antlers, enlarged muscles, gradually become very intolerant of other bulls, and, in autumn, attempt to control and mate with harems of cows. At this time there is much threatening and fighting between rival bulls.[21]

Bull elk, in common with many male mammals live rather solitary lives. The male lynx (_Lynx canadensis_), to take another example, lives alone in vast boreal forests all his adult life, engaged in an unending pursuit of snow shoe hares and other prey, and having very little contact with other lynx except with females during a brief annual rut. The female lynx raises her family alone.[22]

In humans, in contrast, there is no marked seasonal hypertrophy of the gonads: mating occurs throughout the year rather than in a brief period of rut.[23] Men, women and children commonly live in close association throughout the year. It is apparent that if the organization of reproductive processes in humans resembled that of elk or lynx, complex human society could not have developed. Men would have no interest in women except during the annual rut, no interest in children at any time, and a limited willingness to tolerate the presence of other men. There could be no family life and little or no cooperative effort by groups of men.

It is interesting to imagine what animals like elk or lynx would be like if they had evolved larger brains and high intelligence but retained the same reproductive and social organization as they have at present. It seems unlikely that such animals would ever evolve language since they would have little to say to one another. Ethics and morality would, no doubt, be quite different from the pattern observed in humans. Prohibitions against murder or theft would be quite incomprehensible among animals who have not the slightest interest in maintaining a cohesive social group. Adultery would be a meaningless concept in the absence of long-term pair-bonding or marriage. The idea of honoring one's father would also appear very strange among animals who do not know who their father is and who could not have a concept approximating the human idea of fatherhood. One comes to the conclusion that human ethics and morality arise directly from the structure of our instinctive tendencies to form social groups. To a large extent, human nature determines human culture. It is, perhaps, worth adding that if the brain circuits that give rise to ethical behavior are developed under genetic control, their development would be expected to vary from one individual to another. Common observation supports this idea. Certain individuals (psychopaths, sociopaths) seem to have little or no appreciation of normal ethical and moral standards. An intelligent lynx might well consider a male human psychopath to be much saner and more reasonable than the general run of humanity.

It has been proposed that human social organization evolved as a result of the selective advantage of group hunting of large prey animals that could not be attacked successfully by a single hunter. This is essentially a theory that human social organization arose in response to the same selection pressures that induced wolves to live in packs.[24] It is also possible that co-operation among male humans evolved to promote success in wars between rival human groups. Organized intergroup conflict has a very long history in our species and also occurs in chimpanzees and in wolves.[25]

Human groups generally have a leader, a chief, chairman, general, president, king, etc., who possesses to an unusual degree the social skills required to make others obey him. It may be assumed that what we may call followership evolved when those who followed an effective leader experienced greater reproductive success than those who did not. The tendency to follow a charismatic leader is perhaps, frequently beneficial, but as in the case of the oystercatcher tricked into incubating a large artificial egg, it is an instinctive tendency which can lead to maladaptive consequences. Consider the millions of soldiers who enthusiastically obeyed the behests of such leaders as Napoleon, Hitler, or Stalin even though it led them to their deaths.

The tendency of humans to invent and follow religious doctrines may be related to the tendency to follow a charismatic leader. Such major religions as Buddhism, Christianity and Islam owe their origin to the remarkable leadership qualities of individual men (Siddhartha Gautama, Jesus Christ, and Muhammad). Present day religious cults generally appear to be dominated by a single charismatic leader, usually male.[26] Religion is a phenomenon that appears to be specific to the human species: religious doctrines would appear to be impossible in animals lacking propositional language. All known human societies, however, from stone-age hunter-gatherers to the inhabitants of wealthy modern nations, have religious practices of some sort. Estimates of the total number of distinct religions throughout history are as high as 100,000.[27] An accurate estimate is impossible to achieve owing to the difficulty of defining religion. The belief systems associated with Nazism or Communism resembled religion in many ways, including adherence to dogmas that cannot be questioned, a worshipful attitude towards the leadership, plus a high degree of intolerance and self-righteousness.

Whether the tendency toward religion is beneficial to mankind or whether it is a harmful side effect of the evolution of human social behavior is open to debate. On the positive side, there is no doubt that membership in a religious community brings security and comfort to many people. Further, Christianity, for example, has inspired much beautiful music and art. On the negative side, there is a long history of human sacrifice, religious wars, crusades, jihads, genocides, and witch burnings inspired by religion plus the increased danger of religious fanaticism in a world armed with modern weapons.[28] It is sobering to consider that the organized genocides occurring in Nazi Germany, Eastern Europe, the former Yugoslavia, Rwanda, and Cambodia in the twentieth century could not have occurred in any species other than our own.

The long sad history of torture, murder and war have suggested a long-standing question: are humans good by nature or are they evil? William Shakespeare[29] tells us through the words of Hamlet, "What a piece of work is man! How noble in reason, how infinite in faculty, in form and moving how express and admirable, in action how like an angel, in apprehension how like a god – the beauty of the world, the paragon of animals!" Charles Darwin, living during the Victorian period, a time of relative peace, prosperity and continual improvement in many aspects of life, thought that "of all the differences between man and the lower animals, the moral sense or conscience is by far the most important."[30] Those now living, who have witnessed (or at least learned of) the horrific genocidal wars, the torture, murder, and rape occurring in Europe, Africa, South America, and Asia during the past 100 years, may possibly be excused for suspecting that Shakespeare and Darwin, despite their genius, may have been misled by the optimism of the historical periods in which they lived. Many today would be more inclined to agree with Jonathan Swift[31] who described mankind as Yahoos, animals of a nature cunning, malicious, treacherous, cowardly, insolent, libidinous, abject and cruel. Perhaps Karl Linnaeus[32] misjudged the case when he conferred the Latin name *Homo sapiens* (wise man) on humankind. It could be argued that man possesses not wisdom but only a kind of low cunning. Perhaps a better name than *Homo sapiens* would be *Homo vafer* (sly, cunning, or crafty man).

From the point of view of evolutionary biology it makes little sense to view actions as moral or immoral. The important question is: are the actions adaptive? Do they contribute to survival and reproductive success? Humans are animals adapted to live by hunting and gathering in small family groups. Although conflicts do inevitably occur in such groups, people tend to be kind toward relatives and friends but, as recent history teaches, are easily aroused to display the utmost cruelty toward humans outside what is perceived as the family group. From this perspective, humans cannot be described as good or evil: they are simply animals attempting to live their lives in a world which differs greatly from the world to which they were originally adapted.

Notes

1. White, G. (1994). *The natural history and antiquities of Selborne*, London: The Folio Society (first published in 1788).
2. Lorenz, K.Z. (1981). *The foundations of ethology*. (Translated from German by K.Z. Lorenz and R.W. Kickert) New York: Springer-Verlag. Tinbergen, N. (1951). *The study of instinct*. New York: Oxford University Press.
   Tinbergen, N. (1953). *The herring gull's world*. London: Collins. The term "releasing stimulus" seems to have been adopted as a result of the observation that even a weak sensory input may elicit a strong motor reaction, together with the theory that motor reactions were actively inhibited prior to their "release". In general, a historically-minded neuroscientist will note that some of Lorenz and Tinbergen's central concepts are merely a restatement of the concepts of Sherringtonian reflexology. Thus, the terms "eliciting stimulus" or "adequate stimulus" seem preferable to "releasing stimulus" since: (a) they have historical priority; and (b) do not refer to an unsupported theory. Further, the lack of dependence of rather complex behavior patterns on sensory feedback was first demonstrated by Sherrington [Sherrington, C.S. (1906). Observations on the scratch-reflex in the spinal dog. *Journal of Physiology, 34*: 1–50] who noted that rhythmical scratching movements in a spinal dog are not impaired by surgical section of the dorsal root (sensory) fibers supplying the active limb. This constitutes an early demonstration of a spinal pattern generator.
3. Although these ideas appear to be very plausible, it is not easy to demonstrate rigorously that they are correct. A review by U.M. Savalli (The evolution of bird coloration and plumage elaboration: A review of hypotheses, in: Power, D.M. (ed.) *Current ornithology*, 1995, *12*: 141–190) discusses various theories offered to account for the varieties of color and appearance in different species of birds.
4. Beautifully illustrated introductions to the relations between behavior and various human anatomical specializations have been published by: Morris, D. (1977). *Man watching: a field guide to human behavior*, London: Jonathan Cape; and Morris, D. (1985). *Body watching: a field guide to the human species,* London: Jonathan Cape.
5. Lehrman, D.S. (1970). Semantic and conceptual issues in the nature-nurture problem. Pp. 17–52 in: L.R. Aronson, E. Tobach, D.S. Lehrman, and J. Rosenblatt (eds.) *Development and evolution of behavior: Essays in memory of T.C. Schneirla*, San Francisco: W.H. Freeman.
6. Caro, T.M., and Hauser, M.D. (1992). Is there teaching in nonhuman animals? *Quarterly Review of Biology, 67*: 151–174.
7. Kuo, Z.Y. (1930). The genesis of the cat's response to the rat. *Journal of Comparative Psychology, 11*: 1–35.
8. Roberts, W.W. (1970). Hypothalamic mechanisms for motivational and species-typical behavior. In: Whalen, R.E., Thompson, R.F., Verzeano, M. and Weinberger, N. *The neural control of behavior*. New York: Academic Press, pp. 175–206.
9. Goodenough, F.L. (1932). Expression of the emotions in a blind-deaf child. *Journal of Abnormal and Social Psychology*, 27: 328–333.
   Thompson, J. (1941). Development of facial expression in blind and seeing children. *Archives of Psychology*, #264: 1–47.
10. Fulcher, J.S. (1942). "Voluntary" facial expression in blind and seeing children. *Archives of Psychology*, #272: 1–49.

11. The most comprehensive textbook on comparative neuroanatomy in English, though now badly out of date, is: Kappers, C.U., Huber, G.C., and Crosby, E.C. (1965). *The comparative anatomy of the nervous system of vertebrates, including man, vols 1–3*, New York: Hafner Publishing Co. (first published 1936). A more recent textbook is: Butler, A.B., and Hodos, W. (1996). *Comparative vertebrate neuroanatomy: Evolution and adaptation*, New York: Wiley-Liss. An excellent textbook focussed mainly on the human brain but containing some data on other species is: Brodal, A. (1981). *Neurological anatomy in relation to clinical medicine*, 3rd ed., New York: Oxford University Press.

12. Diamond, M.C. (1988). *Enriching heredity: the impact of the environment on the anatomy of the brain*. New York: The Free Press.
   Kolb, B. (1995). *Brain plasticity and behavior*. Mahwah, New Jersey: Lawrence Erlbaum Associates.

13. Anderson, S.R., and Lightfoot, D.W. (1999). The human language faculty as an organ. *Annual Review of Physiology, 62*: 697–722.
   Pinker, S. (1994). *The language instinct*. New York: William Morrow and Company.

14. Damasio, H. (1991). Neuroanatomical correlates of the aphasias. In: Sarno, M.T. (ed) *Acquired aphasia*, 2nd ed. San Diego: Academic Press, pp. 45–71.

15. Myers, R.E. (1976). Comparative neurology of vocalization and speech: proof of a dichotomy. *Annals of the New York Academy of Sciences, 280*: 745–757.

16. Penfield, W., and Roberts, L. (1966). *Speech and brain mechanisms*, New York: Atheneum.

17. Jurgens, U., and Ploog, D. (1970). Cerebral representation of vocalization in the squirrel monkey. *Experimental Brain Research, 10*: 532–554.
   Robinson, B.W. (1967). Vocalization evoked from forebrain in *Macaca mulatta*. *Physiology and Behavior, 2*: 345–354.

18. Jensen-Jazbutis, G.T. (1971). Clinical-anatomical study of microcephalia vera (a microcephalic brother and sister with atrophy of the left mammillary body) *Zeitschrift für Hirnforschung, 12*: 287–305.
   Ross, J.J., and Frias, J.L. (1977). Microcephaly. In: P.J. Vinken and G.W. Bruyn (eds) in collaboration with N.C. Myrianthopoulos, Congenital malformations of the brain and skull, Part I, volume 30, pp. 507–524, *Handbook of clinical neurology*, Amsterdam: North-Holland Publishing Company. A classical description of the behavior of a microcephalic woman can be found in: Korsakov, S.S. (1956). On the psychology of microcephalics. *American Journal of Mental Deficiency, 62*: 108–121 (first published 1894).

19. Skinner, B.F. (1974). *About behaviorism*, New York: Alfred A. Knopf.

20. Breland, K., and Breland, M. (1966). *Animal behavior*, New York: Macmillan.

21. Geist, V. (1982). Adaptive behavioral strategies. In: Thomas, J.W., and Toweil, D.E. (eds) *Elk of North America: ecology and management*, Harrisburg, Pennsylvania: Stackpole Books and the United States Department of Agriculture, Forest Service, pp. 219–277.

22. Ewer, R.F. (1973). *The carnivores*, Ithaca, New York: Cornell University Press.

23. It is interesting that although humans cannot generally be regarded as seasonal breeders there are seasonal variations in gonadal hormone levels and a statistical tendency for more births to occur in the spring than in other seasons in northern countries such as Sweden and Finland [Lam, D.A., and Miron, J.A. (1994). Global patterns of seasonal variation in human fertility, *Annals of the New York Academy of Sciences, 709*: 9–28]. One may, perhaps, regard this as a vestigial piece of physiology and behavior dating back to a time when human primate ancestors gave birth in spring and the young matured sufficiently in a single summer to better withstand the trials of winter.

24. Sanford, C.B. (1999). *The hunting apes: meat eating and the origins of human behavior.* Princeton, New Jersey: Princeton University Press. Schaller, G.B., and Lowther, G.R. (1969). The relevance of carnivore behavior to the study of early hominids. *Southwestern Journal of Anthropology, 25*: 307–341.
25. Dyer, G. (2004). *War,* Canada: Random House.
    Goodall, J. (1986). *The chimpanzees of Gombe: patterns of behavior.* Cambridge: Harvard University Press.
26. Barrett, D.V. (2001). *The new believers: a survey of sects, cults, and alternative religions,* London: Cassell and Co.
27. Wilson, E.D. (1978). *On human nature,* Cambridge, Mass: Harvard University Press, p. 169.
28. Harris, S. (2004). *The end of faith: religion, terror and the future of reason,* New York: W.W. Norton and Company.
    Russell, B. (1961). *The basic writings of Bertrand Russell.* By R.E. Egner and L.E. Dononn (eds.), London: George Allen and Unwin, Ltd. For Russell's comments on religion see pp. 73–99 and 565–604.
29. Greenblatt, S. (ed., 1997). *The Norton Shakespeare,* New York: W.W. Norton and Co. The quotation is from Hamlet, Act 2, Scene 2.
30. Darwin, C. (1998). *The descent of man,* New York: Prometheus Books (first published in New York in 1874). The quotation is taken from p. 100.
31. Swift, J. (1970). *Gulliver's travels,* 2nd ed. (first published in 1727). See Part IV: *A voyage to the country of the Houyhnhnms.*
32. Singer, C. (1931). *A history of biology,* 3rd ed., London: Abelard-Schuman.

VI. Memory and experience-dependent behavior

If one considers critically the hypothesis that the brain is organized in terms of conventional psychological processes, the concept of memory provides a major test case.[1] Our language attaches great importance to memory. We commonly say that people may have a good memory or a bad memory, that someone forgot (failed to remember) when they neglected to do something that had been expected of them, that a senile individual has lost his memory, etc. Furthermore, there has been an enormous amount of psychological and neuroscientific research on memory.

The assumption that memory has a distinct localization in the brain has a very long history. Nemesius, Bishop of Emesa in Turkey, proposed, about 390 AD, that the lateral ventricles of the brain received sensations and generated imagination, that the third ventricle housed cognition and reason, and that the fourth ventricle housed memory.[2] These concepts concerning the brain persisted throughout the medieval period and into Renaissance times, but increasing knowledge of the brain eventually led to the belief that neural tissues, rather than the fluid-filled ventricles, were the essential substrates of function. The psychological component of these concepts remained essentially unaltered however, and present-day ideas that "memory" is located in the hippocampus, the thalamus, the frontal lobe, etc., are clearly a modern reworking of 1600-year old ideas.

What are we really talking about when we use such words as "memory" and "learning?" These words can be defined in at least three different ways. (1) In psychological terms, memory is a mental process distinct from other mental processes such as sensation, perception, attention, motivation, or emotion. "Memory" in this sense usually refers to a conscious experience, a recollection of some event or scene; "learning" is the process of establishing such a memory. (2) In behavioral terms, learning and memory refer to the establishment and maintainance of long-lasting experience-dependent adaptive changes in behavior occurring within the lifetime of one individual. (3) In neuroscientific terms, learning and memory refer to the establishment and maintainance of experience-dependent changes in synaptic transmission in the nervous system. Much confusion has been generated in the study of memory by a failure to make clear distinctions between these different types of definition.

A further point of major importance is that much of the basic brain circuitry underlying behavior is established by developmental processes which are little affected by varying individual experience. In the laboratory rat, for example, it is generally understood that the patterns of locomotion, feeding behavior, grooming behavior, mating behavior, maternal behavior, etc., are instinctive in this sense.

There is much evidence that learned behavior is a refinement and further elaboration of instinctive behavior. For this reason, animals have great difficulty acquiring behaviors for which they have no instinctive predisposition. For example, the instinctive pack hunting of dogs permits easy co-operation with humans in joint attacks on prey but the instinctive solitary hunting of cats does not permit this. Dogs, but not cats, will point out game to a hunter, retrieve shot small game, and co-operate with humans in the pursuit of larger game such as deer, bear, wild boar, large cats, etc. Similarly, humans normally learn to speak but other animals do not because they lack the basic circuitry underlying linguistic abilities which develops spontaneously in the normal human brain (see Chapter V).

It appears that instinctive and learned (individually acquired) behaviors share a common neural basis. For example, in the laboratory rat, destruction of areas of the neocortex or hippocampus that impair learned behavior, e.g. running through a maze without making errors, will also impair the performance of instinctive behaviors, e.g. maternal behavior, sexual behavior, or hoarding food. Similarly, in the intact brain, the large-scale patterns of electrical activity of the hippocampus and neocortex during learned behavior are essentially the same as the large-scale patterns of electrical activity of these structures during instinctive behavior.[3] These facts are consistent with behavioral evidence suggesting that learned behavior is a refinement and extension of instinctive behavior. Therefore, in an adult laboratory animal or in a human, performance of a learned behavior will inevitably be associated with the activation of neural circuitry that evolved to control instinctive behavior.

Bearing these points in mind, let us consider how the three types of definition of memory, the psychological, the behavioral, and the neural, relate to neuroscientific data. A simple behavioral definition of memory does not appear to be adequate for the analytic work required in neuroscientific studies. Consider, for example, the definition of learning offered in a popular textbook of animal behavior "the durable modification of behavior in response to information acquired from specific experiences". If an experimentally produced brain lesion, for example, were found to abolish a learned visual discrimination, we might doubt that the effect was due to a loss of "memory". It might be attributable to an interruption of visual input to the brain, to impaired motoric abilities, or other factors. If good performance in a test of learning and memory is really dependent on the innate brain circuitry involved in instinctive behavior, then a brain lesion that removes such circuitry will impair performance even though "memory" was not directly affected. We recognize intuitively that correct performance of a learned behavior depends upon the normal functioning of a multiplicity of systems and that a "memory", an "engram", or "neuroplasticity" will be only one component among many. In a normal working brain in an animal engaged in some behavior or other, there will always be a wide array of different processes all operating at the same time. Some of these processes will be part of the innate sensori-motor circuitry involved in performing the learned behavior but there will be others that have only an incidental relation to the acquired performance. A definition of memory that is to be useful in neuroscience must take this into account and devise adequate means of distinguishing between "memory" and the other activities that are simultaneously in progress with it during the course of behavior.

It is surprising that these basic points have often been neglected in experiments in which measures of the electrical or chemical activity of the brain have been studied in relation to performance in tests of learning. The behavioral tests of "learning and memory" used in such work have varied widely, ranging form eyelid conditioning through training in various types of mazes, shock avoidance tests, and delayed response or delayed matching tests, all in laboratory animals, to memory for stories, nonsense syllables, etc., in humans. Although the details of these tests are largely irrelevant to the point to be made here, it is of great importance that recording only some arbitrarily defined aspect of behavior as "the measure of learning" ignores much of what is actually going on during a behavioral test. Rats in a maze or a Skinner box (a box in which a rat can press a lever to obtain food according to some more or less complicated schedule) do not make only the responses being recorded (errors and correct choices in the maze, lever presses in the Skinner box). They walk or run about at varying speeds, sniff at the floor, rear, pause to face-wash or scratch themselves, or stand stock-still for varying periods. In correlation with these varying behaviors there will be changes in respiration, heart rate, blood pressure, and in core temperature. Not only are all these activities likely to change in a systematic way during the course of training or retention performance in the learning task, but they are also associated with distinctive patterns of brain activity. Some of these patterns are antecedent to specific behaviors and may play a role in causing them; others are consequent to specific behaviors because they result from behavior-dependent sensory feedback.

Therefore, if one finds systematic changes in some type of brain activity during the course of training in a behavioral task, at least two types of conclusions are possible. (a) The brain activity in question is directly related to learning and memory processes. (b) The brain activity in question is related in some way to behavior or physiological processes which are changing during the course of the experiment but may not be recorded systematically by the experimenters and may have nothing directly to do with "memory". The history of electrophysiological studies of brain activity during tests of learning and memory contains many instances in which a pattern of electrical activity (spontaneous field potentials, artificially evoked potentials, spontaneous or evoked unit discharges) which was at first heralded as a sign of memory was subsequently found to be related to gross motor activity, olfactory input elicited by sniffing, or changes in core temperature. Motor activity, time spent sniffing, and core temperature all change during the course of a behavioral learning experiment and brain activity will change in correlation with these phenomena. However, since identical phenomena occur during spontaneous behavior when no training is involved, there is no justification for assuming that any of them have anything directly to do with "memory".

Similar considerations apply to studies in which brain imaging techniques are used in conjunction with an experiment on "memory" in humans. During, the course of a memory experiment there are likely to be systematic changes in muscle tone, minor movements (fidgeting), respiration, etc. Any changes in brain activity which are detected in different phases of the experiment may be related to such factors rather than to memory itself. Such possibilities must be systematically investigated.

A popular method of studying the neural basis of "memory" has consisted of examining the effect of localized destruction of various parts of the brain on an animal's ability to perform in a behavioral test. The method is important because it offers a means of achieving a better understanding of the effects of brain injury by disease or trauma in humans but there has been very little tendency to make a rigorous inquiry into the meaning of the results obtained. It is very common to see experimental papers with titles of the general form "Contributions of structure X to learning and memory." The contents of the paper reveal that destruction of structure X (e.g. hippocampus, various parts of the neocortex, thalamus, basal ganglia, etc) impairs an animal's ability to run a maze or perform some other more or less complicated task. What does this really mean? No one supposes that behavioral testing of an animal with a hole somewhere in the brain will tell us anything specific about the details of synaptic change that may result from training. The details of synaptic function and their alteration by individual experience clearly require a very detailed approach at a cellular and physicochemical level. The value of the brain lesion-behavioral-memory-testing method is evidently presumed to lie in its supposed ability to tell us something about the neural basis of memory considered as a psychological category. There are several difficulties with this. First, lesions in all the major brain regions (cerebral cortex, subcortical white matter, basal ganglia, basal forebrain, diencephalon, brainstem, cerebellum) have been found to impair some form or other of individually acquired behavior. Similarly, it is generally accepted that damage to the temporal lobes, frontal lobes, thalamus, basal forebrain and subcortical white matter can all produce losses of "memory" in humans. Consequently "memory", considered as a psychological process, does not have a discrete localization in the brain. The ancient theory of Nemesius that there is a distinct faculty of memory with a discrete localization in the brain is clearly wrong and should be abandoned.

A second problem with attempts to identify a brain location for memory as a psychological process is that although learning and memory have always been thought of as something quite distinct from "instinct", the evidence from research on animals makes it quite clear that learned behaviors are a secondary modification of instinctive behaviors and, further, that the neural modifications produced by individual experience occur within the neural circuitry that forms the basis of instinctive behavior.

The idea that individual experience can modify synaptic connectivity within specific pre-existing sensori-motor circuits is supported by a great deal of neuroscientific evidence.[4] Visual experiences can alter the physical appearance and connections of neurons in the visual cortex and can radically alter the responsivity of such neurons to visual stimuli. Somesthetic experiences have similar effects on neurons in somatosensory cortex. Acrobatic training alters the connectivity of neurons in the cerebellum. The neural changes that are responsible for behaviors acquired as a result of individual experience do not depend on a specialized "memory system" with a circumscribed location in the brain. It is more likely that plastic changes occur concurrently at many loci within the sensori-motor systems activated by the environmental situation in which the learning occurs.

The preoccupation with a circumscribed "memory system" located some-where in the brain has prevented people from appreciating the widespread anatomical and physiological effects of individual experience. Learning and memory are a part of the broad range of processes referred to as "adaptation ". Long practice in riding a bicycle or paddling a canoe produces not only a steady improvement in skill but also a multiplicity of changes throughout the body.[5] The development of calluses on the soles or palms are obvious to everyone. The increased size of muscle cells, the changes in muscle protein levels, the increased thickness of cartilage on the surfaces of joints, the increased size and strength of tendons, and changes in cardiovascular activity, such as a fall in resting heart rate, are less obvious but can all be readily demonstrated by appropriate procedures.

The very structure of the bones of the skeleton is determined to some extent by the forces developed during habitual activities. New bone is laid down along the lines of maximum compression or tension in long bones such as the femur. The upper end of the tibia has a distinct groove caused by pressure from the patellar ligament in people who habitually sit in a squatting position but non-squatting people have no trace of such a groove.[6] Similarly, one must expect that there will be many adjustments made throughout the body of a habitual scholar who spends much time sitting quietly at a desk. The changes occurring in the central nervous system as a result of experience should be viewed as components in the overall process of adaptation.

A further implication of the findings indicating that learning is a secondary modification of brain circuits involved in instinctive behavior is related to conventional classifications of "memory". It has long been thought that conditioning and learning includes a number of distinct categories including habituation, classical or Pavlovian conditioning, and operant conditioning, plus numerous categories of "memory" including; imaginal memory, associative memory, episodic memory, semantic memory, procedural memory, declarative memory, iconic memory, short-term memory, long-term memory, working memory, reference memory, verbal memory, spatial memory, evaluative memory, autobiographical memory, automatic memory, effortful memory, etc. Apart from the fact that these terms seem to spring up in an undisciplined manner, like weeds in an untended garden, it is apparent that they refer to an overall behavioral performance involving complex sensori-motor circuitry. The performance of intact or brain-injured humans or laboratory animals on a complex behavioral test cannot provide specific information concerning any neuroplastic elements that may be involved. The behavioral performance is due to the output of a complex neural system including both plastic and non-plastic elements. It is conceivable that the plastic changes are actually quite similar in all cases and that it is the properties of the non-plastic neural elements involved in the performances that have given rise to many of the distinctions noted above.

One can conclude that the conventional category of "memory" as a psychological process is not a useful concept in neuroscientific studies. In order to make advances in our understanding of the neural basis of experience dependent changes in behavior, we must adopt a neuroscientific definition of learning and memory. Although there is no doubt that our everyday language will continue to speak of "pleasant memories", "unhappy memories", "losing one's memory", etc., we must remember that the behavioral performances referred to by such terms are dependent on neural circuitry laid down primarily by developmental processes as well as neural circuitry developed as a result of individual experience. A victim of Alzheimer's disease, for example, loses not only learned behavior but instinctive behavior as well.

This does not necessarily mean that the term "memory" should be abandoned. Many outdated concepts survive harmlessly in everyday English and even in scientific terminology. The element "oxygen" is so named in English because it was once believed, erroneously, that it was responsible for acidity (Greek: oxys, sour; plus gennan, to produce). "Sauerstoff", the German word for oxygen, has a similar origin. No one now is troubled or confused by this. We recognise that our language, illogical though it often is, is a cherished product of a long history but, nonetheless, we must take care that it does not interfere with our ability to understand the natural world.

Notes

1. Most of the ideas discussed in this chapter are presented in a more detailed and technical form in: Vanderwolf, C.H. (2001). The hippocampus as an olfacto-motor mechanism: were the classical anatomists right after all? *Behavioural Brain Research, 127*: 25–47; and in Vanderwolf, C.H., and Cain, D.P. (1994). The behavioral neurobiology of learning and memory: a conceptual reorientation. *Brain Research, 19*: 264–297.
2. Marshall, L.H., and Magoun, H.W. (1998). *Discoveries in the human brain.* Totowa, New Jersey: Humana Press.
3. Vanderwolf, C.H. (2003). *An odyssey through the brain, behavior, and the mind.* Boston: Kluwer Academic Publishers.
4. Bailey, C.H., and Kandel, E.R. (1993). Structural changes accompanying memory storage. *Annual Review of Physiology, 55*: 397–426.
   Buonomano, D.V., and Merzenich, M.M. (1998). Cortical plasticity: from synapses to maps. *Annual Review of Neuroscience, 21*: 149–186.
5. Astrand, P.-D., and Rodahl, K. (1970). *Textbook of work physiology,* New York: McGraw-Hill Book Company.
   Harries, M., Williams, C., Stanish, W.D., and Micheli, L.J. (1998). *Oxford textbook of sports medicine,* 2nd edition, Oxford: Oxford University Press, see pp. 301–320, pp. 379–388, and pp. 389–404.
6. Kate, B.R., and Robert, S.L. (1965). Some observations on the upper end of the tibia in squatters, *Journal of Anatomy, 99*: 137–141.

VII. Neural mechanisms of locomotion in humans

It is instructive to consider human locomotion from an evolutionary perspective.[1] The earliest land vertebrates, amphibians and reptiles, arose from fish, animals which swim by making side to side movements of the trunk and tail. This pattern of movement is preserved in four legged reptiles which walk and run by moving diagonal pairs of limbs together (the left foreleg moves with the right hind leg; the right foreleg moves with the left hind leg) assisted by lateral undulations of the trunk and tail. This pattern can be observed in the locomotion of crocodiles, alligators and quadrupedal lizards. Some reptiles (e.g. the basilisk, a type of tropical American water lizard) run bipedally using alternating movements of the hind legs. Mammals evolved an entirely new form of locomotion, the gallop, in which the trunk is flexed in a dorso-ventral direction, rather than laterally, and the hind limbs are moved forward in synchrony (approximately) and in alternation with the forelegs which are also moved forward in approximate synchrony. Fast bipedal locomotion in mammals other than humans seems to consist of hopping forward on the hind legs, a pattern that can be thought of as a bipedal form of the gallop. This type of locomotion is observed in the various types of kangaroos of Australasia, jerboas (desert rodents of Africa and Asia), the kangaroo rats of the Western hemisphere, the springhaas of South Africa, and the elephant shrew of Africa. All these animals have long hind legs, a long tail, and a hopping or bouncing type of locomotion. The long tail is not essential however since the arctic hare can hop rapidly on its long hind legs, with its body erect, even though it has only a very short tail.

Humans locomote almost exclusively by means of alternating movements of the hind limbs. Hopping forward, with both legs moving forward together, is for us a slow, unnatural, and effortful means of progression. This is consistent with the idea that humans evolved from a line of arboreal climbing mammals rather than terrestrial galloping mammals.

During walking or running, humans move the arms together with the legs in a pattern that preserves the ancestral diagonal pattern of locomotion; the left arm swings forward with the right leg and the right arm swings forward with the left leg. The arm movements are produced by active muscular contractions which persist even when the arms are tied to the trunk to prevent swinging.[2]

This suggests that the arm and leg movements are co-ordinated by a pattern generator in the central nervous system.

The anatomical locomotion of the pattern generators for human locomotion is not well understood. Unlike the situation in spinal dogs and cats, clear instances of stepping have seldom been observed in spinal humans, suggesting that the evolution of the unique form of locomotion observed in humans was associated with a reduction or loss of the ancestral spinal pattern generators for stepping. Spinal monkeys also do not display stepping under the same conditions in which stepping is easily elicited in spinal cats or dogs.[3]

Furthermore, although high decerebrate monkeys display typical decerebrate rigidity, a posture in which all four limbs are extended, locomotor patterns have not been observed. If only the neocortex is removed in an experimental animal (decorticated preparation) locomotion on a level surface is excellent in rats or cats, as we have seen (see Chapter III) and is also present (though rather feeble) in monkeys if care is taken to prevent the development of muscle contractures.[4]

If the forebrain is extensively damaged in humans as a result of accidental mechanical trauma or the growth of large intracranial tumors, a decerebrate posture is assumed in which the legs are stiffly extended. The arms may also be stiffly extended, resembling the situation in laboratory animals, but in some cases the arms are held in a semi-flexed posture. These different postures may be due to differences in the parts of the brain that are damaged.[5] In macaque monkeys a transection through the lower midbrain results in decerebrate rigidity with extension of all four limbs while removal of the cerebral cortex alone tends to produce, at times, a posture with extension of the hind limbs and flexion of the upper limbs. Despite this, decorticate monkeys are able to walk, as already noted. In human cases of decortication or decerebration (the extent of damage is often difficult to ascertain with precision), the ability to walk appears to be lost. Passive rotation of the head to the right usually elicits extension in the right arm and increased flexion in the left arm; passive rotation of the head to the left usually elicits extension of the left arm and flexion of the right arm. These reactions are brainstem reflexes closely resembling those elicitable in decerebrate cats or monkeys.

A related posture often occurs in hemiplegia, a disorder usually occurring in humans as a result of a hemorrhage or plugging (due to an embolus or to atherosclerotic narrowing) of the middle cerebral artery or its branches. This may result in the destruction of a large area of the neocortex or of its descending connections in the internal capsule on one side of the brain, leading to a flaccid paralysis of the leg and arm on the side opposite (contralateral) to the brain injury (see Chapter VIII, The neural control of voluntary movement in humans). With the passage of time the tonus of the muscles in the affected limbs increases, resulting in a characteristic standing posture in which the leg is stiffly extended while the arm is flexed against the chest. This posture is comparable to the extension in all four limbs observed in decerebrate quadrupeds if one considers that the anti-gravity muscles in an erect human include the extensors in the legs and the flexors in the arms.[6]

The paralyzed arm and hand will assume an extensor posture in some hemiplegic patients if they are placed in a quadrupedal attitude, standing on the floor with the hands resting on the seat of a chair. The fingers are held in a flexed posture with the weight of the body supported by the knuckles, a posture which closely resembles the normal standing posture of chimpanzees and gorillas. These observations suggest that pattern generators for knuckle standing were present in early human ancestors and further, that they still persist in the modern human brain but are normally suppressed by more recently acquired neural circuitry. After a cerebral lesion the ancient coordination pattern may reveal its existence.[7]

Locomotion is very much impaired in hemiplegia, partly because the patient has little or no voluntary control of the leg, especially in the early flaccid paralysis stage of hemiplegia. Laboratory animals, such as rats, do not display flaccid paralysis after removal of neocortex and can walk, climb and support their weight by holding on to the edge of a vertical board within a few hours after their surgery. In the later stages of human hemiplegia, locomotion is still impaired because the development of strong extensor muscle tone makes it difficult to flex the knee and hip. The patient may be able to walk by swinging the stiffly extended leg out in a wide arc, scraping the toe and medial side of the foot on the ground as he does so. When one compares this situation to the excellent locomotor abilities of decorticate or high decerebrate rats and cats, and also to the presence of quadrupedal locomotion in decorticate monkeys, one is led to the conclusion that the evolution of bipedal locomotion in man may have been associated with a transfer of the pattern generators for locomotion from spinal and brain stem mechanisms to mechanisms located in the neocortex and structures associated with it.

The evolutionary transition from quadrupedal to bipedal locomotion re-quired the solution of several complex physical problems. A walking quad-rupedal ape has a low center of gravity and the body is always supported by at least two legs on opposite sides of the body. In contrast, an erect human has a high center of gravity and must support the weight of the body on only one leg whenever a step is taken. This position is mechanically unstable. A normal human standing erect on two legs has a center of gravity located above a point midway between the two feet. Before one foot can be lifted to take a step, the center of gravity must be shifted to a point over the other foot. Consequently, the upper body must be rocked to the right in order to take a step with the left foot and it must be rocked to the left in order to take a step with the right foot. These lateral rocking movements can be seen with especial clarity in a young child walking or running directly toward or away from the observer because children tend to place their feet rather far apart to improve postural stability.

In addition to rocking from side to side, bipedal forward locomotion requires that the body be inclined forward so that the center of gravity is located over a point anterior to the toes. During a step, the body begins to fall diagonally forward and to the side of the lifted leg. If a human is filmed while walking past a wall inscribed with a grid of lines, it can be seen that the head bobs down and up again with each step. Turning is accomplished by allowing the body to fall slightly toward the desired side. Walking backwards is accomplished by leaning backwards slightly, the reverse of walking forward.

Since people lean forward at a slight angle whenever they are walking forward, the center of gravity, even at its highest, is lower than it is during quiet standing. It has been pointed out that if a human were to walk through a tunnel with a height exactly equal to the height during immobile standing, there would be a centimetre or more of clearance between the top of the head and the tunnel during walking.

Our understanding of the relation between brain activity and the physical requirements of bipedal locomotion was greatly improved by observations made by J.P. Martin in a group of patients who had developed Parkinson's disease following inflammation of the brain (encephalitis) brought on by influenza during the great world-wide influenza epidemic of 1918–1920.[8] Martin showed that many of the difficulties in locomotion that are characteristic of Parkinson's disease are due to a loss of the lateral rocking and forward leaning postural reactions which are essential for human locomotion. Some patients who could stand erect quite well were unable to walk, the feet appearing to "stick to the ground". However, since normal stepping could be elicited if someone else, walking behind, rocked their upper body from side to side, it appears that the stepping mechanism was quite normal. What was lacking was the ability to rock the body from side to side. Other patients, lacking the ability to initiate and maintain forward leaning of the body, could walk only if someone else held their upper body in a forward leaning posture. Some of the patients with this disability had discovered for themselves that they could walk if they held a weight, such as a chair, before them to force the body to lean forward. Still others were able to initiate walking but could not stop because they were unable to straighten up from a forward leaning position. Since the leaning forward posture in such patients typically increased progressively once it was initiated, the patients were forced to walk more and more rapidly, then run uncontrollably forward until they fell or collided with some object. Uncontrollable forward locomotion (festination), a rather common symptom in Parkinson's disease, therefore, appears to be due to a lack of control over the posture of leaning forward. Since individual patients could lose either the rocking or forward leaning abilities, or both together, the two seem to be dependent on separable mechanisms.

Patients who have lost their locomotor abilities as a result of Parkinson's disease appear to have returned to a condition similar to that present in very young infants. If a new-born child is held gently around the trunk with the feet in contact with a horizontal surface, stepping movements can usually be elicited by: (a) holding the body in a forward leaning position; (b) moving the body forward while (c) rocking the upper part of the body from side to side.[9] This reflexive stepping, which can be elicited on a vertical surface or even on the ceiling, normally disappears before genuine spontaneous locomotion occurs.

If we could determine the precise brain structures which, when damaged in Parkinson's disease, give rise to the peculiar locomotor disorder characteristic of this condition, our understanding of human locomotion would be greatly increased. Parkinson's disease is generally regarded as a disease of the basal ganglia, a term which refers primarily to the substantia nigra in the midbrain, the striatum (including the caudate nucleus and the putamen) and the pallidum (often known as the globus pallidus). Cases of post-encephalitic Parkinson's disease studied by J.P. Martin had suffered an extensive loss of neurons in both the striatum and pallidum. However, many of the symptoms of Parkinson's disease may also occur after isolated destruction of a class of dopamine-containing neurons in the substantia nigra which project to the striatum.[10] It is often the case that destruction of a variety of different components of a complex working mechanism give rise to essentially similar symptoms. The basal ganglia have complex anatomical connections with the neocortex, parts of the thalamus (the lateral and anterior parts of the ventral thalamic nuclei) and with the tectum (superior colliculus). These structures undoubtedly also interact with the cerebellum via corticopontocerebellar pathways. The cerebellum projects to: the reticular formation and vestibular nuclei (influencing reticulospinal and vestibulospinal pathways; the red nucleus (influencing rubrospinal pathways); to the lateral and anterior parts of the ventral thalamic nuclei which project to the sensori-motor areas of the neocortex; and to the intralaminar nuclei of the thalamus which project to the caudate nucleus and the putamen. Exactly how this complex neural circuitry controls locomotion has yet to be determined.

Notes

1. Smith, J.M. (1966). *The theory of evolution*. Harmondsworth, Middlesex, England: Penguin Books Ltd.
2. Fernandez Ballesteros, M.L., Buchthal, F., and Rosenfalck, P. (1965). The pattern of muscular activity during the arm swing of natural walking. *Acta Physiologica Scandinavica*, *63*: 296–310.

A diagonal pattern of limb movement is characteristic of primates in general. See: Larson, S.G. (1998). Unique aspects of quadrupedal locomotion in nonhuman primates, In: Strasser, E., Fleagle, J., Rosenberger, A., and McHenry, H. (editors) *Primate locomotion: Recent advances*, New York: Plenum Press, pp. 157–173.

3. Eidelberg, E. (1981). Locomotor control in macaque monkeys. *Brain, 104*: 647–663.
4. Denny-Brown, D. (1966). *The cerebral control of movement*. Springfield, Illinois: Charles C. Thomas, see p. 90.

     Travis, A.M., and Woolsey, C.N. (1956). Motor performance of monkeys after bilateral partial or total cerebral decortications. *American Journal of Physical Medicine, 35*: 273–310.
5. Fulton, J.F. (1949). *Physiology of the nervous system*, 3rd ed. New York; Oxford University Press.
6. In South American sloths, animals which spend their lives clinging to the underside of branches, the antigravity muscles of both the limbs and the trunk are flexors. Consequently decerebrate rigidity in the sloth consists of a posture in which the neck and trunk are ventroflexed and the limbs are flexed against the body. This indicates that in decerebrate rigidity antigravity muscles are strongly contracted regardless of whether they are flexors or extensors. See: Richter, C.P., and Bartemeier, L.H. (1926). Decerebrate rigidity of the sloth. *Brain, 49*: 207–225.
7. Brain, W.R. (1927). On the significance of the flexor posture of the upper limb in hemiplegia, with an account of a quadrupedal extensor reflex. *Brain: 50*: 113–137.
8. Martin, J.P. (1967). *The basal ganglia and posture*. London: Pitman Medical Publishing Company Limited.
9. Peiper, A. (1963). *Cerebral function in infancy and childhood*. New York: Consultants Bureau, The International Behavioral Sciences Series.
10. Schultz, W. (1982). Depletion of dopamine in the striatum as an experimental model of Parkinsonism: direct effects and adaptive mechanisms. *Progress in Neurobiology, 18*: 121–166.

VIII. The neural control of voluntary movement in humans

Many of our current ideas about the cerebral control of movement are based on observations of the loss of voluntary movement in the unfortunate human victims of stroke. Before discussing this topic in more detail it is important to consider a fundamental question: what is a voluntary movement?

In everyday speech and according to the definitions supplied by dictionaries, a voluntary action is one produced by an act of the will. Such a definition is not very helpful however, since it merely replaces the original question by one even more difficult. What is the will?

Voluntary control of movement is associated, for many people, with the concept of free will. What is free will? We are accustomed to say that people are free to act as they will if their action is not constrained by external forces or threats. This commonsense idea is partly similar to the scientific concept of spontaneous activity. An isolated heart placed in a warm solution containing certain amounts of the chlorides of sodium, potassium, and calcium, plus glucose and oxygen, will beat rhythmically for an indefinite period. This activity is due to the properties of special pacemaker cells, located in the sino-atrial node, which permit rhythmical fluxes of ions through the cell membrane. Pacemaker cells also exist in the nervous system. Humans normally wake up and go to sleep at fairly regular intervals, even when they are living in deep caves in an environment of utter darkness with no perceptible daily variations in temperature, humidity, etc. It is likely that such a circadian rhythm, as it is known, is due to the activity of special pacemaker cells located in the suprachiasmatic nucleus in the hypothalamus.[1] Spontaneous actions, then, can be regarded as originating from causes entirely within an animal, especially within the nervous system, as opposed to those that originate from external causes such as sensory stimuli. Under normal circumstances, of course, behavior is always a joint result of the spontaneous activity of the nervous system interacting with the effects of sensory inputs.[2]

A commonsense view of free will may also include the idea that our action is not determined: we have the feeling that in any situation we could have done something different from what we actually did. It is, however, unlikely that any behavior is truly undetermined, i.e. random. Anyone whose behavior was truly random and completely unpredictable would be astonishing. Most human behavior is readily predictable, as pointed out many years ago by David Hume[3] in some famous passages. "There is a general course of nature in human actions, as well as in the operations of the sun and climate." "Are the changes of our body from infancy to old age more regular and certain then those of our mind and conduct?" "Is it more certain that two flat pieces of marble will unite together than that two young savages of different sexes will copulate? Do the children arrive from this copulation more uniformly than does the parent's care for their safety and preservation?" Hence, it may be said that although human behavior may, in principle, be completely predictable it is only partially predictable in practical terms because it is impossible to have a complete knowledge of all the relevant causal factors.

It is conventional to contrast voluntary actions with reflex actions which occur promptly and automatically in response to an adequate stimulus. For example, such actions as sneezing, coughing, shivering, vomiting, or a startle response to a sudden loud sound are elicited by definite stimuli and are difficult or impossible to suppress by voluntary effort. Actions of this type are also difficult or impossible to perform upon request although this is, of course, very easy to do in the case of fully voluntary acts. Yawning, laughing, and weeping are behavior patterns which are not reflexive in any simple sense but are also not fully voluntary since they may be difficult to suppress in socially inconvenient circumstances and difficult (for most people) to produce upon request.

The concept of operant behavior developed by E.L. Thorndike and B.F. Skinner corresponds in many ways to the common sense idea of a voluntary action.[4] An operant behavior is one whose future probability of occurrence is modifiable by its consequences. Thus, in the terminology of operant conditioning, an action which is followed by the occurrence of a positive reinforcement, such as the delivery of food for a hungry animal, is likely to be repeated, while an action which is followed by the occurrence of a negative reinforcement, such as the delivery of a noxious stimulus (e.g. an electric shock) is less likely to be repeated. It is worth considering the methods that have been developed. Training a hungry rat or a dog for example, is begun by presenting an effective exteroceptive stimulus such as a loud click immediately before the presentation of a small piece of food. After a number of such pairings the auditory stimulus acquires two properties: (a) it serves as a signal to the animal to go to the place where food is presented; and (b) it becomes a secondary or conditioned reinforcer capable of increasing the probability of occurrence of preceding behavior. This conditioned reinforcer can then be used to establish a desired behavior by a series of successive approximations ("shaping" behavior). This means that initially, spontaneous behaviors of approximately the desired type are reinforced but as training proceeds, the requirements for reinforcement are made more and more stringent. By using such methods, many different species have been taught to perform varied and complex actions involving head movement, locomotion, and the manipulation of objects. Pigeons can be taught to play ping-pong; sea lions can be taught to play baseball, and so forth.

Some behavior patterns, however, are not amendable to operant conditioning. That is, the frequency of their occurrence in the future cannot be altered systematically be making the delivery of reinforcement contingent on their occurrence.[5] Making the delivery of a conditioned reinforcer plus food contingent on the occurrence of face-washing does not increase the probability of face-washing in rats. Other behaviours such as yawning or the pelvic thrusting movements of copulation are also not available as operants in a variety of mammals. Presumably, such movement patterns are similar to involuntary movement patterns in humans.

It appears that some behaviors, the various forms of locomotion for example, can be readily brought into the service of diverse functions including feeding, predator avoidance, reproduction, and temperature regulation while other behavior patterns may be closely linked to a single functional system making their occurrence unlikely when the relevant functional system is not activated. Face-washing in a rat for example, seems to be largely dedicated to care of the skin, shivering is largely restricted to situations in which the body temperature has declined appreciably. Smiling and laughing seem to occur in humans largely or entirely in specific social situations. People seldom laugh when they are alone. In contrast, such movement patterns as turning the head, locomotion, and the manipulation of objects can occur at virtually any time and can be used to obtain food, put on a sweater on a cold day, write a letter, make a phone call to a friend, etc. Such behaviors can be thought of as voluntary or operant.

Certain reflexive behavior patterns are initiated by a voluntary act. In the initial phase of swallowing, for example, a voluntary movement of the tongue moves material from the mouth into the pharynx. This triggers a reflex response activating some twenty muscles in overlapping sequence over a period of about 500 milliseconds. A pattern generator co-ordinating all this activity appears to be located in the reticular formation in the medulla.[6] The reflex character of the entire response is revealed by the fact that swallowing is impossible if the initial afferent input is blocked by painting the pharyngeal surface with cocaine.[7] A similar conclusion is suggested by the common observation that swallowing is difficult or impossible if the mouth is dry and empty; in this case there is no stimulus to elicit reflex swallowing.

In the foregoing example, a voluntary act sets the stage for the activation of a reflex response which is in itself not under direct voluntary control. This type of organization of motor acts is probably quite common. One can imagine that the onset of locomotion involves creating a situation in which spinal locomotor reflexes are activated. Similarly, voluntary sexual mounting may set the stage for the activation of spinal copulatory reflexes.

The conclusion that some behaviors are not fully voluntary or are not readily available as operants may appear to be contradicted by the common observation that good actors can produce convincing displays of laughter, weeping, etc., upon request. However, there is good evidence from human stroke patients that voluntary smiling, laughing, weeping, etc., depend on neural mechanisms different from those involved in the unfeigned forms of these behaviors.

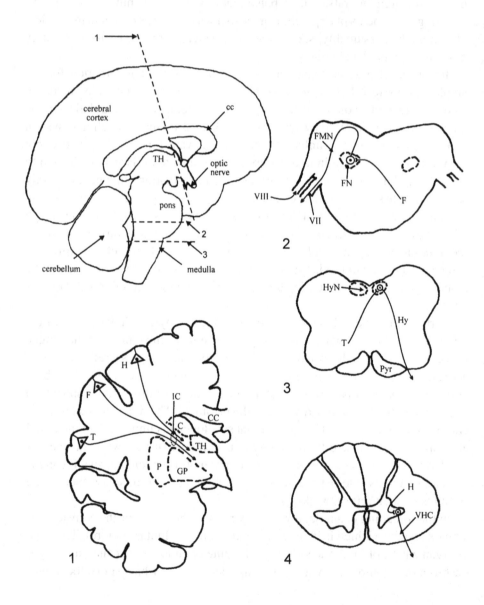

It is unfortunately an all too common occurrence for the blood supply to the internal capsule and part of the basal ganglia to be interrupted by atherosclerotic narrowing and plugging of the middle cerebral artery or some of its branches in the human brain. The blood supply may also be interrupted by an embolus (for example, a blood clot originating in some distant site) or by hemorrhage following the rupture of an artery. In all such cases, the end result is that a region of nervous tissue dies, often producing a behavioral syndrome known as stroke.[8] If a brain lesion interrupts large numbers of axons in the internal capsule, a common occurrence in stroke, the ability of the neocortex to control motor activity is diminished or lost (Figure VIII.1).

A severe stroke typically results in a complete paralysis (hemiplegia) and a reduction or disappearance of muscle tonus on the contralateral (opposite) side of the body. The affected limbs hang limply and cannot be moved voluntarily by the patient (e.g. when requested to do so). In the weeks and months following a stroke, muscle tonus gradually recovers and may become greater than normal. Reflexes are at first reduced or abolished, but they recover after a few days and eventually tend to become more vigorous on the paralyzed side of the body than on the normal side. Thus, stretch reflexes[9] such as the knee jerk reflex are likely to be abnormally vigorous in a hemiplegic limb. High muscle tone in combination with exaggerated stretch reflexes is known as spasticity.

Figure VIII.1. Some descending motor pathways from the human neocortex. *Top left*: A crude sketch of a human brain divided along the anatomical midline. CC, corpus callosum, a major fibre tract connecting the right and left neocortex; TH thalamus. The dotted lines numbered 1–3 refer to the cross sections shown in the remainder of the figure. 1. A cross section through one half of a human forebrain. C, caudate nucleus; CC, corpus callosum; F, layer V pyramidal cell in the face area; GP, globus pallidus; H, layer V pyramidal cell in the hand area; IC, internal capsule, a large fiber pathway including a wide variety of ascending sensory and descending motor pathways; P, putamen; T, layer V pyramidal cell in the tongue area; TH, thalamus. 2. A cross section through the pons in the human brain. F, descending axon from pyramidal cell F; FN, facial nucleus; FMN, facial motor neuron, VII, facial (seventh) cranial nerve; VIII, statoacoustic (eighth) cranial nerve, carrying auditory and vestibular inputs. 3. A cross section through the medulla in a human brain. Hy, hypoglossal motor neuron; HyN, hypoglossal nucleus; Pyr, pyramidal tract carrying descending corticospinal fibers; T, descending axon from pyramidal cell T. 4. A cross section through the cervical part of a human spinal cord. H, descending axon from pyramidal cell H which has followed a pathway through the pyramidal tract and the lateral spinal column. Note that the axons of pyramidal cells F, H, and T cross (decussate) to the opposite side of the brain to influence motor neurons connected to muscles on the opposite side of the body. As a result, destruction of cortex or of descending fibers in the internal capsule will produce a loss of voluntary control of body parts contralateral to the lesion but reflex functions may survive. Thus, a patient who cannot stick his tongue out voluntarily may retain a normal ability to lick water from the lips after taking a drink.

In intact humans, a tactile stimulus moving along the outer margin of the sole of the foot in heel-to-toe direction elicits curling of the toes (ventroflexion or plantar flexion reflex) but in a hemiplegic limb the toes fan out and the big toe extends (dorsiflexion). This reaction, widely known as a Babinski reflex, has long been regarded as indicative of destruction of descending cortical pathways for voluntary movement.

The symptoms of hemiplegia are reduced somewhat with the passage of time. The degree of recovery of function varies with the size and location of the lesion and the amount of training and exercise undertaken by the patient. It is presumed that intact motor pathways can, to some extent, take over the function of those that were lost. There may even be an apparent complete recovery after a small lesion but deficits tend to persist indefinitely with large lesions. Among the earliest movements to recover are co-ordinated flexor synergies of an entire limb so that flexion of the fingers, for example, occurs in association with flexion of the wrist, elbow and shoulder.[10] Somewhat later, an extensor synergy may appear as a co-ordinated extension of the digits, the wrist, the elbow and the shoulder. Isolated movements of a single joint or a single digit recover much later, if at all. The gross flexor and extensor synergies are of little practical use to the patient. It might be thought that a global extensor synergy might be used to reach out for a glass of water which could then be raised to the mouth, or at least near it, by means of a flexor synergy. The flexor synergy, however, seems to be inevitably associated with pronation of the hand (turning the palm downward) so that the glass is emptied before it nears the mouth.[11]

It is apparent that destruction of the neocortical motor areas or their efferent pathways in the internal capsule has a much more severe effect in humans than in laboratory animals such as rats or cats (see Chapter III). This suggest that the evolution in humans of bipedal locomotion and use of the forelimbs primarily for the manipulation and carrying of objects, involved a reorganization of brain motor patterns, giving the neocortex a primary role in controlling the new behavior patterns. However, severely spastic and hemiplegic limbs are not always totally immobilized. Numerous "associated movements" may occur after a period of recovery. Some of these are clearly components of the brainstem reflexes, observed in decerebrate states. Thus, turning the head strongly toward the paralyzed limb may cause it to become extended: turning the head strongly toward the sound limb may cause the paralyzed limb to flex. Clenching the fist of the sound upper limb may produce movements in a paralyzed upper limb but the actual pattern of such movement tends to vary from one patient to another. Spontaneous yawning and stretching movements are usually associated with extension of a paralyzed upper limb.[12]

Some movements retain a normal pattern of co-ordinated activity on the hemiplegic side of the body and the amplitude and rate of the movements may be normal or even greater than normal. A hemiplegic patient may display a reduction in the movement of the chest wall contralateral to the brain lesion when asked to draw a deep breath but a normal or even greater than normal movement of the chest wall contralateral to the brain lesion when breathing spontaneously. This demonstrates a reduction in voluntary control of the respiratory muscles coupled with a normal or an exaggerated automatic or reflexive control of the same muscles.[13] Some patients may be unable to draw a breath or hold their breath voluntarily when asked to do so even though their spontaneous breathing is quite normal, and it is clear that they understood what was asked.

Much the same situation can be demonstrated in the control of facial expression. A hemiplegic patient may have an obvious impairment in the ability to show the teeth voluntarily on the side opposite to the brain lesion when requested to do so, together with an exaggerated display of the teeth on that side when smiling or laughing occur naturally in a social situation.[14] In cases in which the cortical motor pathways are damaged bilaterally, vigorous laughter or weeping with abundant facial movement may occur spontaneously or in response to slight stimuli, even though voluntary movement of the face and limbs is virtually abolished.[15]

Similarly, voluntary swallowing may be lost in a patient whose reflexive swallowing is quite normal. Perhaps what is lost in the latter case is primarily voluntary control of the tongue. Some patients who cannot stick the tongue out voluntarily may, nonetheless, be capable of licking water from their lips in a perfectly normal manner after drinking.

If the body temperature is reduced to elicit shivering, the resulting tremors are of normal amplitude or of greater than normal amplitude in a hemiplegic limb as compared to a contralateral sound limb.[16]

One may conclude that stroke-induced hemiplegia is associated with a reduction in purely voluntary control of the contralateral limbs together with normal or even exaggerated contralateral movement in the case of motor patterns that are not fully voluntary in an intact human. This category includes breathing, smiling, laughing, weeping, yawning, stretching, shivering, swallowing, associated postural reactions, and perhaps other behaviors that have not yet been adequately studied.

Many patients with Parkinson's disease display a behavioral syndrome which is, in certain respects, the opposite of the syndrome observed in hemiplegic patients. Spontaneous laughter, smiling, etc., may be absent giving the patients a poker-faced appearance even though voluntary control of the facial muscles appears to be unimpaired. Similarly, the ability to hold the head erect, an automatic behavior in normal people when they are awake, may disappear in Parkinson's disease, allowing the head to slump forward on to the chest when the patients are sitting quietly awake. Despite this, such patients may be able to extend the head forcefully against an examiner's hand when requested to do so.[17]

The foregoing clinical neurological observations indicate that phasic voluntary movements are dependent on neural circuits different from those involved in more automatic or reflexive behaviors including those concerned with involuntary postural reactions. Consequently, an actor who can feign genuine laughter, weeping, etc., is making use of brain systems different from those involved in the natural occurrence of these behaviors.

Clinical observations of this type have traditionally been interpreted as indicating that motor control in humans is dependent on two distinct systems. (1) Pyramidal tract projections consisting of corticospinal fibers originating primarily in the sensori-motor cortical areas have been supposed to be responsible for all voluntary movement. (2) An extrapyramidal system consisting of the basal ganglia and brain stem projections to the spinal cord have been supposed to be responsible for various non-voluntary and postural reactions. More recently this concept has lost favor, partly because section of the pyramidal tract results in a loss of discrete movements of the extremities rather than a loss of voluntary movement as a whole. Furthermore, lesions apparently restricted to the caudate nucleus and putamen (with secondary degenerative changes in other closely connected structures) have been reported to produce a loss of voluntary control of the fingers and toes without affecting voluntary control of proximal joints (shoulder, hip) in a human patient.[18] This is similar to the deficits reported following section of corticospinal fibers. These clinical observations have been supported by animal experiments showing that neuronal destruction restricted to either the caudate nucleus and putamen or the ascending dopaminergic projections from the substantia nigra to the caudate nucleus and putamen produce impairments in rats using a forelimb to reach for food.[19] These impairments are of the same general type as those produced by destruction of the sensori-motor cortex or section of the pyramidal tract. One can only conclude, as in the case of the problem of the neural central of locomotion, that discrete movements of the limbs are produced by complex neural circuits involving many brain structures. The "associated movements" and other automatic movements observed in hemiplegic patients are presumably the result of activity in surviving neocortex, the basal ganglia, the brainstem, and the cerebellum both ipsilateral and contralateral to the brain lesion.

A traditional view of the sensory control of voluntary movement held that conscious sensations occurred in the primary sensory areas (such as the striate area in the occipital lobe in the case of vision, or Heschl's gyrus in the temporal lobe in the case of audition). Impulses transmitted transcortically to the adjoining association cortex then led to the elaboration of perceptions and ideas which in turn might activate the motor areas (primarily the precentral gyrus) transcortically to produce a voluntary movement. However, this simple view is unlikely to be correct. W. Penfield, who performed surgical removals of various neocortical areas for the treatment of epilepsy, showed that removal of the gyri surrounding the precentral motor region did not prevent dextrous voluntary movement.[20] Therefore, the transcortical excitation theory of motor control must be incorrect. Penfield suggested that the motor cortex must normally be activated and controlled by ascending impulses from systems in the brainstem and diencephalon. Experiments on animals indicate a similar conclusion, as discussed in a previous essay (Chapter III).

In attempting to understand neural control of movement, it is important to be aware that even the simplest of movements involve a complex array of motor activities. For example, if a motionless human, standing erect, reaches out with one hand to take a book from a shelf, the change in weight distribution would cause the body to fall forward were it not for a counteracting backward movement of the upper part of the body. W.R. Hess[21] distinguished such general postural adjustments, which he termed the ereismatic phase of support, from the final directed or telokinetic phase of the total movement (the reaching movement of the arm and hand). When one considers the behavioral complexity of even a rather simple movement, it is not surprising that many different brain structures collaborate in the performance of any natural behavior. In order to understand how the brain controls any such natural behavior, it is clear that very detailed analytical studies of the behavior are essential.

Notes

1. Moore-Ede, M.C., Sulzman, F.M., and Fuller, C.A. (1982). _The clocks that time us_, Cambridge, Massachusetts: Harvard University Press.

2. In the nineteenth and early twentieth centuries, it was assumed by many scientists that behavior was entirely reflexive, i.e. determined by sensory inputs. The conditioned reflex theory of I.P. Pavlov was firmly based on this concept [Pavlov, I.P. (1960). _Conditioned reflexes_, New York: Dover Publications, Inc. (first published, 1927)]. The discovery that the brain displays unceasing activity, even under conditions of minimal sensory input, was revolutionary. However, some students of behavior were already well aware that much normal behavior is spontaneous and independent of any specific sensory input. Thus, the concept of operant conditioning assumes that animals generate behavior spontaneously and that the environmental effects of a behavior (beneficial or noxious) will then influence the probability of its future occurrence (see text).

3. Hume, D. (1978). _A treatise of human nature_, 2nd edition (edited by L.A. Selby-Bigge and P.H. Nidditch), Oxford: Clarendon Press (first published 1739–40). Pp. 401–402.

4. A good recent textbook on operant conditioning and learning in general is: Domjan, M. (1998). _The principles of learning and behaviour_, 4th edition. Pacific Grove, California: Brooks/Cole Publishing Company.
   A simple and practical discussion on the training of animals is provided by: Skinner, B.F. (1951). How to teach animals. _Scientific American, 185_: 26–29.

5. Annable, A. and Weardon, J.H. (1979). Grooming movements as operants in rats. *Journal of Experimental Analysis of Behavior, 32*: 297–304.
   Breland, K., and Breland, M. (1966). *Animal behavior.* New York: The Macmillan Company.
   Shettleworth, S.J. (1975). Reinforcement and the organization of behavior in golden hamsters: hunger, environment and food reinforcement. *Journal of Experimental Psychology: Animal Behavior Processes, 1*: 56–87.
6. Doty, R.W., Richmond, W.H., and Sorey, A.T. (1967). Effects of medullary lesions on coordination of deglutition. *Experimental Neurology, 17*: 91–106.
7. Pommerenke, W.T. (1928). A study of the sensory areas eliciting the swallowing reflex. *American Journal of Physiology, 84*: 36–41.
8. A good general textbook on hemiplegia and other neurological conditions is: Walton, J. (1993). *Brain's diseases of the nervous system,* 10th edition. Oxford: Oxford University Press.
9. Sudden stretching of a muscle stimulates specialized intramuscular receptor organs (muscle spindles) which activate sensory fibers projecting to the spinal cord. These fibers in turn excite motor neurons that project to the stretched muscle causing it to contract briefly. A well-known stretch reflex is the knee-jerk elicited by tapping on the patellar tendon (just below the knee cap) while the lower leg hangs freely in a sitting subject. The blow on the tendon causes a sudden stretching of a large muscle (quadriceps) on the anterior surface of the thigh, thereby eliciting a quick reflex contraction of that muscle.
10. Twitchell, T.E. (1951). The restoration of motor function following hemiplegia in man. *Brain, 74*: 443–480.
11. Brodal, A. (1973). Self-observations and neuroanatomical considerations after a stroke. *Brain, 96*: 675–694.
12. Walshe, F.M.R. (1923). On certain tonic or postural reflexes in hemiplegia with special reference to the so-called "associated movements". *Brain, 46*: 1–37.
13. Jackson, J.H. (1899). Case of left hemiplegia with turning of the eyes to the right-slightly greater amplitude of the left side of the chest in inspiration proper and slightly less amplitude of movement of that side in voluntary expansion of the chest. *Lancet, XIX*: 1659–1660.
    Cohen, E., Mier, A., Heywood, P., Murphy, K., Boultbee, J., and Guz, A. (1994). Diaphragmatic movement in hemiplegic patients measured by ultrasonography, *Thorax, 49*: 890–895.
    Prezedborski, S., Brunko, E., Hubert, M., Mavroudakis, N., and Zegers de Beyl, D. (1988). The effect of acute hemiplegia on intercostal muscle activity, *Neurology, 38*: 1882–1884.
    Simon, R.P. (2001). Breathing and the nervous system. In: Aminoff, M.J. (editor) *Neurology and general medicine,* New York: Churchill Livingstone, pp. 1–21.
14. Monrad-Krohn, G.H. (1939). On facial dissociation. *Acta Psychiatrica et Neurologica, 14*: 557–566.
15. Ironside, R. (1956). Disorders of laughter due to brain lesions. *Brain, 79*: 589–609.
16. Uprus, V., Gaylor, G.B., and Carmichael, E.A. (1935). Shivering: a clinical study with especial references to the afferent and efferent pathways. *Brain, 58*: 220–232.
17. Martin, J.P. (1967). *The basal ganglia and posture.* London: Pitman Medical Publishing Co.
18. Oppenheimer, D.R. (1967). A case of striatal hemiplegia. *Journal of Neurology Neurosurgery and Psychiatry, 30*: 134–139.
19. Whishaw, I.Q., O'Connor, W.T., and Dunnett, S.B. (1986). The contributions of motor cortex, nigrostriatal dopamine and caudate-putamen to skilled forelimb use in the rat. *Brain, 109*: 805–843.
20. Penfield, W. (1954). Mechanisms of voluntary movement. *Brain, 77*: 1–17.
21. Hess, W.R. (1954). *Diencephalon: autonomic and extrapyramidal functions.* New York: Grune and Stratton. (see p. 31).

IX. About hunting

According to Gilbert White, an eighteenth century English clergyman and naturalist, "there is such an inherent spirit for hunting in human nature, as scarce any inhibitions can restrain."[1] The view that hunting and meat eating are at least partially instinctive in humans is consistent with archaeological evidence that human ancestors have been hunters for at least 2.5 million years (see Chapter IV). Clearly, hunting is not simply an arbitrary custom dependent on the existence of a particular culture. Nonetheless, many people today regard hunting as a cruel, barbaric form of behavior which is inconsistent with the ethics of modern civilization.[2] What do we know about hunting and how should we regard it?

Predation in the natural world. Walking alone one evening on a quiet gravel road near where I live in southern Ontario, I heard the repeated cries of a bird originating high up in a large maple tree. Although the sun was setting, the light was sufficient for a clear view of a raccoon standing on a branch directly in front of a large hole in the tree trunk. I could see the raccoon's head and jaws moving and I could hear the sounds of its munching interspersed with the diminishing cries of the dying bird on which it was evidently feeding.[3]

I have witnessed scenes of predation involving various species on a number of occasions in my life. They were always disturbing. One wishes to punish, kill, or at least drive off the predator. However, on longer reflection, one cannot avoid the fact that predation occurs everywhere in the natural world, not only among mammals and birds, but everywhere throughout the animal kingdom. Predation appears to play an essential role in regulating the abundance of many species. Many features of animals that we consider beautiful such as the slim legs and elegant bounding gait of white tailed deer, the thunderous take-off of ruffed grouse, the heavy protective shell of turtles, the quills of porcupines, as well as such qualities as the prodigious reproductive capacity of rats and mice, probably owe their very existence to the selective effects of predation. Animals that were able to avoid predators by high speed flight or by means of protective defences, as well as those that were able to offset the losses due to predation by high fertility, have been more likely to leave surviving descendants than those who lacked these features. It is clear that predation has played a major role in shaping the natural world.

It is also clear that the natural world was not designed according to the moral and ethical principles by which humans, sometimes successfully, attempt to regulate their interactions with one another. One can understand the conclusions of some religious groups (Gnostics, Manicheans, Zoroastrians) that the world we see around us could not possibly be the work of a kind, beneficent god and must, therefore, have been constructed by an evil being, a devil who, perhaps, accomplished his wicked design in a moment when the kind god was napping.

Hunting ethics. Hunting and the eating of meat require the killing of animals. This fact has troubled thoughtful people, not only in our own time, but also in earlier periods. The historian John Cummins[4] tells us that some medieval ecclesiastics, living in a time when hunting was highly esteemed, nonetheless believed that hunters were wicked. Jean-Jacques Rousseau, an eighteenth century French philosopher, published a long diatribe against the eating of meat.[5] Nonetheless, both hunting and meat-eating have survived, sometimes even in countries in which the official religion forbids the taking of life.[6] Hunters themselves everywhere seem to regard hunting, not as an evil, but as a joyous activity. Edward of Norwich, who wrote the oldest English book on hunting between 1406 and 1413,[7] began a section on the hunting life with the statement "Now shall I prove how hunters live in this world more joyfully than any other men."

It is probable that a very ancient form of justification for hunting was one of necessity. In the words of an Abenaki hunter, presumably addressing the spirit of a newly slain animal, "I have killed you because I need your skin for my coat and your flesh for my food. I have nothing else to live on."[8] Well-fed contemporary hunters living in wealthy western countries, who can make no such claims, sometimes say that their actions are an attempt to restore the balance of nature. There is truth in this. For example, it is generally believed that white-tailed deer populations in North America, no longer controlled by non-human predators, would increase to unsustainable levels if human hunting were to cease.[9] However, it must be acknowledged that men do not take up hunting because they wish to perform a public service. Men hunt because they derive a deep satisfaction and pleasure from the entire process, from the initial pursuit of game to the final sitting down with one's family to an excellent dinner. The fact that many people love to hunt but that there is very little interest in the job of executioner in a slaughterhouse demonstrates that love of hunting is not simply due to "bloodlust" as the critics of hunting have sometimes contended. Human enjoyment of a complex activity cannot be reduced to any single factor. Our enjoyment of a gourmet meal, for example, is due not only to the food or the wine, but also to the pleasure of conversation with our companions, the furnishings of the dining room, etc. Similarly, hunters enjoy being in the woods , enjoy carrying and using a treasured weapon, enjoy the craft of outwitting a wary game animal, enjoy the excitement of the kill and are delighted when they have been successful.

Concern by hunters for the animals they hunt has led to the development of a hunting ethic among hunting peoples everywhere in the world. Although modern hunting organizations have sometimes published ethical guidelines for hunting,[10] these tend to be of a rather generalized nature, avoiding detailed statements as to exactly what a hunter should or should not do. However, such ethical guidelines may encourage: making quick humane kills; making every reasonable effort to follow up and kill any animals that are unintentionally wounded; avoiding the wastage of meat or useful skins; continuing efforts by hunters to increase their knowledge and skill in woodcraft and marksmanship; conservation of game and preservation of natural habitats; and respect for game laws and the rights of private landowners.

It is important to realize that the hunting ethics of traditional hunter-gatherer peoples may have differed substantially from those of modern sport hunters. For example a modern sportsman would not think of killing a female duck accompanied by a family of downy ducklings but an aboriginal hunter who has hungry mouths to feed may not be able to afford such scruples. Aboriginal hunters everywhere have made use of various types of traps (pit-traps, deadfalls, snares, etc.) to secure game but modern sportsmen avoid such methods.[11] In many jurisdictions, in fact, the use of such methods is illegal except in the case of the trapping of fur-bearing animals by a licensed trapper.

Hunters as individuals tend to have a deep affection for the woods and wild creatures in their neighbourhood. One of the essential qualifications of a successful hunter is a practical knowledge of animal ecology and behavior, a knowledge of where various species are to be found and the time of day when they are likely to be active. It is also essential for a hunter to be a keen observer, able to detect well-camouflaged animals hidden in the vegetation. Thus, hunters tend to be practical naturalists who take a naturalist's delight in wildlife and wild places. This often develops into a desire to promote the conservation of wildlife and wildlife habitats. Hunter organizations such as Ducks Unlimited, the Ruffed Grouse Society, and the Ontario Federation of Anglers and Hunters, for example, have been in the forefront of the conservation movement for decades and have been very effective in encouraging the preservation or restoration of wildlife habitats and in the reintroduction of native species in areas in which they had been nearly or completely exterminated. The successful restoration of wild turkeys and Canada geese in Ontario provide examples of this.

It is unclear whether concern for conservation of wildlife is a modern development or whether it is a feature of the hunting ethic in hunter-gatherer societies. For example, although it is currently fashionable to believe that aboriginal North Americans traditionally followed wise conservation practices, there is very little evidence to support this belief. According to Ian McTaggart-Cowan, "There is no evidence that the native people had any concept of numbers applied to their food animals. They took what they could without concern for replacement rate or overkill."[12] However, there is the intriguing observation that Paleolithic reindeer hunters, 50,000 years ago, and even earlier, killed mainly adult male deer.[13] Although it is, of course, impossible to know what the intent of these hunters was, the effect of their behavior would be to preserve the maximum number of breeding female deer. It is conceivable that some conservation practices are very ancient indeed.

A discussion of hunting ethics and conservation practices does not address the question which is central for many people. Is the killing of animals for any human purpose ever morally justifiable?

First, if one accepts that predation is an intrinsic feature of the natural world and that humans are a part of the fauna of our planet, then there appears to be no obvious reason why humans should not be predators. If one believes that the human killing of other animals is unethical, there seems to be no way of avoiding the conclusion that the killing of animals by other non-human animals is also unethical, or at least intolerable and should be stopped. It could, of course, be maintained that non-human predators do not understand that what they are doing is wrong and they cannot, therefore, be held responsible for their actions. A mentally incompetent human who kills another human is also judged not to be criminally responsible but is incarcerated anyway (in a hospital for the criminally insane rather than a prison). Thus, a logical consequence of the idea that killing animals is unethical is that, as ethical beings, we should intervene massively in the natural world, attempting to separate predators from their prey. Some philosophers and animal activists have, in fact, advocated taking this step.[14] Apart from the fact that a program to prevent predation in the natural world is technically and economically impossible and would have a great many undesirable consequences, it rests on a philosophical fallacy. Human ethical principles are our own invention, an outgrowth of the evolution of human social behavior. They are not a part of the natural non-human world and it is presumptuous in the extreme to suppose that the whole world should dance to music of our making. We, however, can choose the ethical principles by which we ourselves wish to live. Consequently, ethical and moral standards change somewhat from one historical period to another. Killing other people, ordinarily considered the most heinous of all crimes, is not only not condemned, but actively encouraged during wartime. The sexual mores of ordinary people in Western countries have changed dramatically in the past fifty years. Similarly, hunting and the killing of other animals for our own purposes will continue as long as large numbers of people support such practices. Ultimately, public opinion is the final arbiter of ethical questions.

Neural basis of hunting behavior. In considering this topic, it is essential to be clear about what it is that we hope to explain in neuroscientific terms. If two human hunters, walking together through woods and open areas, detect several deer a long way off, what do they do? They immediately stop walking and are likely to crouch down, concealing themselves behind rocks, bushes, etc. Conversation is reduced to tense whispers as they consider if any of the deer are suitable prey, and if so, how, considering the terrain, wind direction, and the behavior of the deer, they might be able to approach near enough to make a kill. Their behavior does not in the least resemble the red-faced shouting and threats typically seen in two men in a quarrel. Hunting is not like intraspecific aggression and fighting.[15]

Neuroscientific studies confirm that predatory behavior in laboratory animals is distinct from aggressive or defensive displays. Electrical stimulation of a medial and ventromedial region of the hypothalamus in cats elicits a threat or defensive response pattern which includes opening of the mouth, retraction of the lips, hissing, growling, flattening of the ears, piloerection, and striking with a forepaw plus biting directed towards a nearby rat or an inanimate model.[16] The overall pattern is similar or identical to the behavior displayed by a cat toward a rival cat or towards an aggressive dog.[17] A similar behavior pattern can be elicited from the same ventromedial hypothalamic zone in the marsupial opossum, suggesting that it has a long phylogenetic history.

A predatory type of attack which includes approaching the prey object with the body lowered and the neck extended, followed by the use of the forepaws to catch and hold the prey and by severe biting, can also be elicited by localized electrical stimulation of the hypothalamus. Piloerection does not occur. The predatory attack pattern of behavior closely resembles naturally occurring predatory behavior and is more likely to kill or severely wound the prey than the threat-defensive pattern. Predatory attack can be elicited from a hypothalamic zone lateral or dorsolateral to the threat-defensive zone in both cats and opossums. The predatory attack zone extends rostrally into the preoptic area but, according to W.W. Roberts,[16] the defensive threat zone does not. Therefore, different systems of hypothalamic neurons are involved in the two patterns of behavior.

There is evidence that the amygdaloid nuclei, large cellular complexes located in the temporal region, modulate both defensive-threat behavior and predatory attack by means of neural projections to the hypothalamus. Electrical stimulation of the amygdala can elicit defensive-threat behavior and stimulation of different regions of the amygdala may either suppress or facilitate predatory attack elicited by concurrent hypothalamic stimulation.[16] Large bilateral lesions of the amygdala in cats abolish predatory attack on a bird or a mouse.[18] Extensive removal of the neocortex would, no doubt, also abolish effective predatory behavior (although this point has never been specifically tested as far as I know) but this effect would be part of a generalized impairment of behavior (dementia) rather than a specific (or somewhat specific) loss of predatory behavior. Animals with amygdaloid lesions are able to feed themselves and do not suffer from a global impairment of behavior. Large lateral hypothalamic lesions would also abolish predatory behavior, one can assume, because such lesions abolish most forms of purposive behavior.[19]

It appears that predatory attack behavior in laboratory animals depends on specific neural systems located in the hypothalamus and amygdaloid nuclei but also on more widespread systems including the neocortex. The complete behavioral performance would also, of course, be dependent on other brain structures including the brain stem, cerebellum, and the spinal cord.

There appears to be no systematic evidence on the neural basis of hunting behavior in humans. Although it is certainly possible that the neural systems for predatory behavior in laboratory animals are similarly active in the human case, one cannot be confident about this. Typical quadrupedal predatory behavior involves seizing prey and biting, often directed toward the head and neck, but human predatory attacks ordinarily involve the use of tools (ranging from clubs and rocks to a modern rifle) and do not involve biting. Consequently, it may be that human predatory behavior is not fully homologous with predatory behavior in conventional predators and that it has a unique neural basis.

Notes

1. White, G. (1994). *The natural history and antiquities of Selborne*, London: The Folio Society (first published in 1788), see p. 18.
2. The current popular literature attacking or defending hunting is enormous. Two books written from the point of view of thoughtful hunters are: (1) Petersen, D. (editor). *A hunter's heart: honest essays on the blood sport*, New York: Henry Holt and Company, 1997; and (2) Swan, J.A. (1996). *In defense of hunting*, New York: Harper Collins, paperback edition.
3. Many predators kill their prey by eating it. Durward Allen reports the following incident: "On a flight late in the afternoon, Don and Chief Ranger Ben Zerbey saw the moose (this animal had been under observation by wolf investigators using an aircraft) down and the wolf feeding on its rump. The moose lay quietly with its head up watching the wolf," p. 129 in: Allen, D.L. (1979). *Wolves of Minong*, Boston: Houghton Mifflin Company.
4. Cummins, J. (1988). *The art of medieval hunting: the hound and the hawk*, Edison, New Jersey: Castle Books, p. 10.
5. Rousseau, J.J. (1911). *Emile*, London: J.M. Dent and Sons, Ltd. (Translated from French by B. Foxley; first published 1762).
6. Harrar, H. (1953). *Seven years in Tibet*. Leicester: Ulverscroft, (translated from the German by R. Graves). The dominant religion in Tibet, a form of Buddhism, forbids the taking of life even in the case of flies or biting insects. Further, animals may not be deprived of their food. Nonetheless, meat, fish and honey were widely eaten in traditional Tibetan society, and fur garments were worn in winter. Tibetans managed to combine these practices with a stated adherence to their ethical principles by making social outcasts of people who performed,

for pay, the tasks of slaughtering animals, preparing skins for clothing and the collection of honey. Similar hypocrises are not unknown in contemporary Western society.

7. Edward of Norwich (1974). *The master of game*, New York: AMS Press, from an original manuscript dating about 1420 (see p. 8). The greater part of this book is a translation of Gaston Foix's (also known as Gaston Phebus) *Livre de chasse*, an illuminated manuscript available today as: Bise, G. (1978). *Medieval hunting scenes*, Fribourg: Production Liber SA (first published about 1390, translated into modern English by J.P. Tallon). For a philosophical discussion of the place of hunting in human life throughout history and its contribution to human happiness, see: Ortega y Gasset, J. (1972). *Meditations on hunting*, New York: Charles Scribner's sons, (Translation from the Spanish by H.B. Wescott).

8. Cassell, J. and Fiduccia, P. (editors) *The quotable hunter*. New York: The Lyon's Press, 1999, p. 5.

9. A particularly clear illustration of the growth in a population of predator-free white tail deer is provided by the George Reserve, a 464 hectare (1,146 acre) plot of Michigan woodland enclosed by an 11.5 foot deer-proof fence. Six deer released in the reserve in 1928 had increased to possibly 220 deer by 1933. Since the vegetation was being severely damaged by so many deer, their numbers were controlled by hunting until 1966 when the population size was gradually and systematically reduced until only about 10 deer remained in 1975. At this point hunting was stopped. Within five years the population had increased to 212 deer. Since then (1980), the population has again been controlled by hunting. [See: McCullough, D.R. (1984). Lessons from the George Reserve, Michigan, In: Halls, L.K. (editor) *White-tailed deer: ecology and management*, Harrisburg, Pennsylvania: Stackpole Books, pp. 211–242.]

10. A magnificently illustrated book [Blüchel, K.G. (2000). *Game and hunting*, Cologne: Könemann Verlagsgesellschaft mbH] on the history and present practices of hunting in Europe contains a statement of the hunting ethics adopted by the Conseil International de la Chasse et de la Conservation du Gibier (CIC; The International Council of Hunting and the Conservation of Game, see pp. 646–647).

11. An excellent summary of aboriginal methods of trapping game is provided by: Coon, C.S. (1971). *The hunting peoples*, Boston: Little, Brown and Company. Although such methods as snaring game may be illegal in Western countries, they continue to be used by poachers. See, for example: Benson, R. (1985). *Ragnar's ten best traps and a few others that are damn good too*, Boulder, Colorado: Paladin Press. *The US army survival manual, FM 21-76*, New York: Dorset Press, 1994, also gives detailed instructions on primitive methods of securing fish and game and preparing it for food.

12. McTaggart-Cowan, I. (1989). Room at the top? In: Hummel, M. (editor) *Endangered spaces: the future for Canada's wilderness*, Toronto: Key Porter Books Ltd., pp. 249–266. Also see: Krech, S. III (1999). *The ecological Indian: myth and history*. New York: W.W. Norton and Company.

13. Gaudzinski, S., and Roebroeks, W. (2000). Adults only. Reindeer hunting at the Middle Palaeolithic site Salzgitter Lebenstedt, Northern Germany. *Journal of Human Evolution*, *38*: 497–521.

14. See: Causey, A.S. (1996). Is hunting ethical? In: Petersen, D. (editor) *A hunter's heart: honest essays on blood sport*. New York: Henry Holt and Company, pp. 80–89.

15. Any simple dichotomy between intraspecific aggression and predatory behavior in humans is complicated by the observation that military snipers and serial killers appear to display an essentially predatory pattern of behavior toward other humans. There appears to be little scientific information available on any of this, however.

16. A general list of references on central elicitation of defensive and predatory behavior includes the following: Hess, W.R. (1957). *The functional organization of the diencephalon*, New York: Grune and Stratton.

Egger, M.D., and Flynn, J.P. (1967). Further studies on the effects of amygdaloid stimulation and ablation on hypothalamically elicited attack behavior in cats, In: W.R. Adey and T. Tokizane (editors) *Structure and function of the limbic system, Progress in Brain Research, 27*: 165–182. Hunsperger, R.W., and Bucher, V.M. (1967). Affective behavior produced by electrical stimulation in the forebrain and brain stem of the cat. In: W.R. Adey and T. Tokizane (editors) *Structure and function of the limbic system, Progress in Brain Research, 27*: 103–127.

Roberts, W.W. (1970). Hypothalamic mechanisms for motivational and species-typical behavior. In: R.E. Whalen, R.F. Thompson, M. Verzeano, and N.M. Weinberger (editors) *The neural control of behavior,* New York: Academic Press, pp. 175–206.

Siegel, A., Roeling, T.A.P., Gregg, T.R., and Kruk, M.R. (1999). Neuropharmacology of brain-stimulation-evoked aggression. *Neuroscience and Biobehavioral Reviews, 23*: 359–389.

17. See: Leyhausen, P. (1979). *Cat behavior: The predatory and social behavior of domestic and wild cats,* New York: Garland STPM Press (translated from the German by B.A. Tonkin).

18. Cherkes, V.A. (1967–1968). Instinctive and conditioned reactions in cats after removal of amygdaloid nuclei, *Neuroscience Translations, 4*: 418–424 (Published by the Federation of American Societies for Experimental Biology for the National Institutes of Mental Health).

19. Levitt, D.R., and Teitelbaum, P. (1975). Somnolence, akinesia, and sensory activation of motivated behavior in the lateral hypothalamic syndrome, *Proceedings of the National Academy of Sciences of the U.S.A., 72*: 2819–2823.

Robinson, T.E., and Whishaw, I.Q. (1974). Effects of posterior hypothalamic lesions on voluntary behaviour and hippocampal electroencephalograms in the rat. *Journal of Comparative and Physiological Psychology, 86*: 768–786.

Index